Materials are an indispensable part of human civilization since the inception of life on earth. With the passage of time, innumerable new materials have been explored as well as developed and the search for new innovative materials continues briskly. Keeping in mind the immense perspectives of various classes of materials, this series aims at providing a comprehensive collection of works across the breadth of materials research at cutting-edge interface of materials science with physics, chemistry, biology and engineering.

This series covers a galaxy of materials ranging from natural materials to nanomaterials. Some of the topics include but not limited to: biological materials, biomimetic materials, ceramics, composites, coatings, functional materials, glasses, inorganic materials, inorganic-organic hybrids, metals, membranes, magnetic materials, manufacturing of materials, nanomaterials, organic materials and pigments to name a few. The series provides most timely and comprehensive information on advanced synthesis, processing, characterization, manufacturing and applications in a broad range of interdisciplinary fields in science, engineering and technology.

This series accepts both authored and edited works, including textbooks, monographs, reference works, and professional books. The books in this series will provide a deep insight into the state-of-art of Materials Horizons and serve students, academic, government and industrial scientists involved in all aspects of materials research.

More information about this series at http://www.springer.com/series/16122

Muhammad Bilal Tahir · Khalid Nadeem Riaz

Nanomaterials and Photocatalysis in Chemistry

Mechanistic and Experimental Approaches

Springer

Muhammad Bilal Tahir
Khwaja Fareed University of Engineering and Information Technology
Punjab, Pakistan

Khalid Nadeem Riaz
Department of Physics
University of Gujrat
Gujrat, Pakistan

ISSN 2524-5384 ISSN 2524-5392 (electronic)
Materials Horizons: From Nature to Nanomaterials
ISBN 978-981-16-0645-8 ISBN 978-981-16-0646-5 (eBook)
https://doi.org/10.1007/978-981-16-0646-5

This Springer imprint is published by the registered company Springer Nature Singapore Pte Ltd.
The registered company address is: 152 Beach Road, #21-01/04 Gateway East, Singapore 189721, Singapore

Contents

About the Authors

Dr. Muhammad Bilal Tahir is a researcher at Khwaja Fareed University of Engineering and Information Technology, Rahim Yar Khan (KFUEIT), Pakistan and currently working on Nanotechnology for Material processing, Photocatalysis, Energy and Environmental applications, Colloids & Surfactants. His initial work was on photocatalysis based Processes for decontamination of pollutants, adsorption of heavy metals in water and oil and energy production. He has 100+ publications in well reputed ISI international journals and serving as Editor in numerous journals. He was the recipient of "BEST TEACHER AWARD 2015–2016" from the University of Gujrat Pakistan on his excellent performance towards teaching and research.

Dr. Khalid Nadeem Riaz is a researcher at the University of Gujrat, Gujrat and currently working on Nanotechnology for Material processing, Photocatalysis, Energy and Environmental applications, Colloids & Surfactants. He completed hid Ph.D. (Material Physics) from Bahauddin Zakariya University, Multan, Pakistan (entire experimental research work done at Zhejiang University, Hangzhou, China. He is currently working as an Assistant Professor at the Department of Physics, University of Gujrat and is involved in different research projects as PI, Co-PI and collaborator. He has also been part of a number of research projects completed in various capacities. He has 90+ publications in a well-reputed ISI international journal.

Chapter 1
Introduction

1.1 Historical Background of Photocatalysis

As early as 1901, the experiments were conducted by chemist Giacomo Ciamician to study the influence of light wavelength on chemical reactions [1]. The performed experiments with red and blue lights demonstrated that a chemical effect can be taken place when only blue light was used in chemical reactions. He carefully explored the probability that whether these chemical reactions were powered by temperature (thermal heat) produced via incident irradiated blue light or any other reason. The term "photocatalysis" was first time used in several scientific studies in 1911. In Germany, Eibner introduced the concept in his report on the influence of ZnO on bleaching of Prussian blue [2–5]. This study has motivated subsequent experiments on ZnO photocatalyst for other reactions including reduction of Ag^+ ion to Ag^o under light illumination in 1924 [6]. Although, prior to these effects, the photo-active chemical reactions had long been investigated which did not use light active catalysts [7]. In 1932, TiO_2 and Nb_2O_5 were studied to investigate the photocatalytic decomposition of $AgNO_3$ to Ag and $AuCl_3$ to Au [8, 9].

Later in 1938, TiO_2 as photosensitizer was studied for decomposing dyes in the oxygen environment [10]. Interestingly, photocatalytic process continued as a main chemical reaction because of the absence of typical commercial applications. However, this situation modified in early 1970 due to two main reasons. Firstly, the oil crises motivated the researchers to search other renewable energy sources besides fossil fuels [11]. Secondly, the serious environmental effects by large-scale industrialization prompted scientists to explore green and efficient energy resources for the future demand of energy in the world [12]. In 1968, O_2 evolution on TiO_2 was first reported by scientists of the Bell Labs [13]. On the other hand, Fujishima and Honda reported H_2 generation from H_2O oxidation via TiO_2 electrodes under UV photo-illumination in 1972 [14]. Similarly, photocatalysis for water splitting for the production of hydrogen (H_2) and oxygen (O_2) under argon atmosphere was reported

M. B. Tahir and K. N. Riaz, *Nanomaterials and Photocatalysis in Chemistry*,
Materials Horizons: From Nature to Nanomaterials,
https://doi.org/10.1007/978-981-16-0646-5_1

in 1977 [15]. The authors observed that production of O_2 and H_2 was greatly inhibited in the presence of nitrogen gas because N_2 was reduced to NH_3 and trace amount of N_2H_2 via TiO_2. During the same era of time, TiO_2, ZnO, and CdS photocatalysts were used for the degradation of CN^- and $SO3^{2-}$ in the presence of light by Frank and Bard [16]. Afterward, Fujishima et al. studied the reduction of CO_2 via photocatalysis utilizing variety of inorganic semiconductor-based photocatalytic materials in 1979 [17]. These early reports increased the photocatalysis applications; therefore, lot of research was done in 1980 using especially TiO_2 nanoparticles as photocatalyst [18–20].

After this, major focus of scientists was on the understanding of fundamental principles, improving performance of photocatalysts, searching new photocatalysts, and expending their applications. For example, in 1997, the light-initiated superhydrophilicity phenomenon was explored using TiO_2. Therefore, TiO_2 was used as building material due to its function as self-cleaner d anti-fog agent [21, 22]. Many new materials having photocatalytic activity higher than TiO_2 were developed, which have higher band gap energies and only function under visible-light illumination [23]. Then, in next step, attempts were made to develop visible-light absorbing materials having higher efficiencies [24, 25]. In parallel, scientists gradually learned more about photocatalysis principles and also extended practical applications of photocatalysis. In conclusion, it can be said that photocatalysis has recently emerged as an advanced and efficient technology, providing simple solutions for environmental and energy crises to our today society.

1.2 Broad Definition of Photocatalysis

The photocatalysis progress was certainly motivated by sunlight-based natural photosynthesis. The photocatalytic process has been extensively used in the literature for two different practices [26]. As shown in Fig. 1.1a, upon absorption of light energy by material, a thermodynamically uphill reaction (Gibbs free energy change (ΔG) > 0) starts usually called photosynthesis. The used material in above process is called "photocatalyst" only if the irradiated light is considered to be a reactant. The major reactions in this category include photocatalytic water splitting and CO_2 conversion [27–29]. On the other hand, if a semiconductor material utilizes light energy for the start of thermodynamically downhill chemical reactions ($\Delta G < 0$, Fig. 1.1b), then in this process material does not alter the reaction thermodynamics but only supports the new reactions via absorption of light. Thus, used material in this reaction would be regarded as true photocatalyst and overall process will be considered as photocatalysis. The phenol oxidation to hydroquinone via oxygen ($\Delta G° = -$ 167.96 kJ/mol) or thorough oxidation to CO_2 and H_2O ($\Delta G° = -$ 3027.36 kJ/mol) are examples of this category [30, 31]. Besides thermodynamic perspective, the photocatalytic properties including absorption of light, charge transport, and separation are shared by both types of reactions (Fig. 1.1).

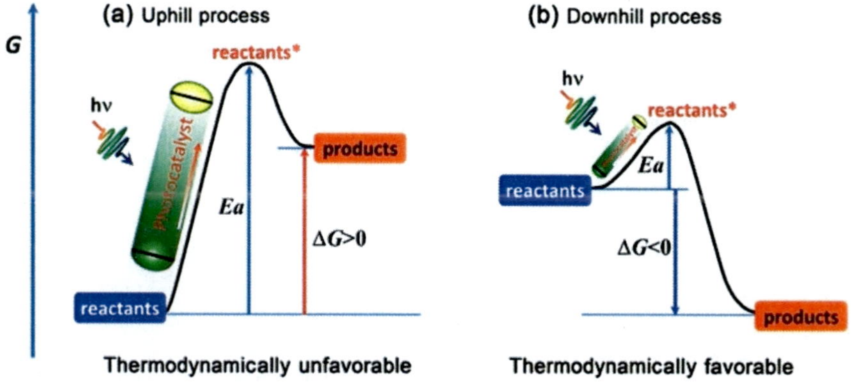

Fig. 1.1 Thermodynamically supported uphill and downhill photocatalysis: **a** uphill process; **b** downhill process. Reproduced from ref [26] with permission from ACS, Copyright 2017

1.3 Classification of Photocatalysis

A promising approach uses solar energy to excite a semiconductor photocatalyst that can enable different redox reactions [32]. This broad field is known as photocatalysis. Each potential redox reaction that can occur signifies an individual photocatalytic field. More specifically, it can efficiently utilize solar energy with simultaneous energy conversion and environmental remediation. The energy conversion aspect majorly includes water splitting, carbon dioxide (CO_2) reduction, nitrogen fixation and synthesis of organic molecules (less explored) to develop sustainable and clean energy. Secondly, its environmental remediation applications are more likely to focus on air purification and wastewater decontamination. The overall photocatalysis details in two aspects: Energy conversion and environmental remediation (Fig. 1.2) will be covered in this book.

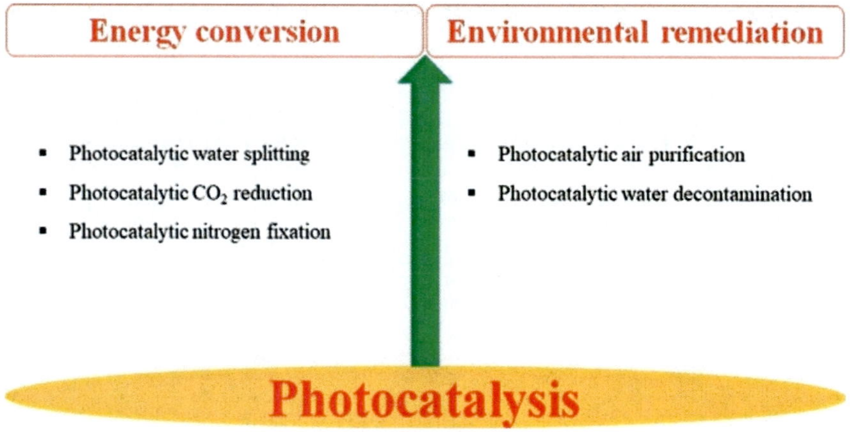

Fig. 1.2 Photocatalysis for energy and environmental applications

References

1. A. Albini, M. Fagnoni, Chemsus chem **1**, 63 (2008)
2. L. Bruner, J. Kozak, Z. Elektrochem, Angew. Phys. Chem. **17**, 354 (1911)
3. A. Eibner, Chem.-Ztg. **35**, 753 (1911)
4. A. Eibner, Chem.-Ztg. **35**, 774 (1911)
5. A. Eibner, Chem.-Ztg. **35**, 786 (1911)
6. E. Baur, A. Perret, Helv. Chim. Acta **7**, 910 (1924)
7. S.E. Braslavsky, M.B. Rubin, Photochem. Photobiol. Sci. **10**, 1515 (2011)
8. C. Renz, Helv. Chim. Acta **15**, 1077 (1932)
9. A. Fujishima, X. Zhang, D.A. Tryk, Surf. Sci. Rep. **63**, 515 (2008)
10. C. Doodeve, J. Kitchener, Trans. Faraday Soc. **34**, 902 (1938)
11. P. Lotfabadi, Renew. Sust. Energ. Rev. **52**, 1340 (2015)
12. D. Ravelli, D. Dondi, M. Fagnoni, A. Albini, Chem. Soc. Rev. **2009**, 38 (1999)
13. P. Boddy, J. Electrochem. Soc. **115**, 199 (1968)
14. A. Fujishima, Nature **238**, 37 (1972)
15. G. Schrauzer, T. Guth, J. Am. Chem. Soc. **99**, 7189 (1977)
16. S.N. Frank, A.J. Bard, J. Phys. Chem. **81**, 1484 (1977)
17. T. Inoue, A. Fujishima, S. Konishi, K. Honda, Nature **277**, 637 (1979)
18. G. Blondeel, A. Harriman, D. Williams, Sol. Energy Mater. **9**, 217 (1983)
19. M. Halmann, M. Ulman, B.A. Blajeni, Sol. Energy **31**, 429 (1983)
20. M. Halmann, V. Katzir, E. Borgarello, J. Kiwi, Sol. Energy Mater. **10**, 85 (1984)
21. R. Wang, K. Hashimoto, A. Fujishima, M. Chikuni, E. Kojima, A. Kitamura, M. Shimohigoshi, T. Watanabe, Nature **388**, 431 (1997)
22. N. Serpone, A. Emeline, S. Horikoshi, V. Kuznetsov, V. Ryabchuk, Photochem. Photobiol. Sci. **11**, 1121 (2012)
23. A. Kudo, K. Sayama, A. Tanaka, K. Asakura, K. Domen, K. Maruya, T. Onishi, J. Catal. **120**, 337 (1989)
24. C. Kormann, D.W. Bahnemann, M.R. Hoffmann, J. Photochem. Photobiol. A: Chem. **48**, 161 (1989)
25. M. Hara, T. Kondo, M. Komoda, S. Ikeda, N.J. Kondo, K. Domen, M. Hara, K. Shinohara, A. Tanaka, Chem. Commun. **3**, 357 (1998)
26. F.E. Osterloh, Photocatalysis versus photosynthesis: a sensitivity analysis of devices for solar energy conversion and chemical transformations. ACS Energy Lett. **2**, 445–453 (2017)
27. M.G. Walter, E.L. Warren, J.R. McKone, S.W. Boettcher, Q. Mi, E.A. Santori, N.S. Lewis, Solar water splitting cells. Chem. Rev. **110**, 6446–6473 (2010)
28. S. Chen, T. Takata, K. Domen, Particulate photocatalysts for overall water splitting. Nat. Rev. Mater. **2**, 17050 (2017)
29. X.X. Chang, T. Wang, J.L. Gong, CO_2 Photo-reduction: insights into CO_2 activation and reaction on surfaces of photocatalysts. Energy Environ. Sci. **9**, 2177–2196 (2016)
30. J.A. Dean, N.A. Lange, *Lange's Handbook of Chemistry*, 15th edn. (McGraw-Hill, New York, 1999)
31. X. Yang, D. Wang, Photocatalysis: From fundamental principles to materials and applications. ACS Appl. Energy Mater. **1**, 6657–6693 (2018)
32. J. Luo, S. Zhang, M. Sun, L. Yang, S. Luo, J.C. Crittenden, A critical review on energy conversion and environmental remediation of photocatalysts with remodeling crystal lattice, surface, and interface. ACS Nano **13**, 9811–9840 (2019)

Chapter 2
Fundamentals of Photocatalysis for Energy Conversion

2.1 Photocatalytic Water Splitting

2.1.1 Importance of Photocatalytic Water Splitting

Energy shortage is the most significant problem for the mankind due to the reason that our current society is built on fossil fuels [1]. This fact has motivated researchers to develop clean and renewable energy technologies. For this purpose, hydrogen (H_2) is regarded as one of the cheapest sources of energy, for example, in fuel cell devices, it is used to produce electricity, water, and heat [2]. However, for the production of hydrogen, a complicated process is involved such as hydrocracking of fossil fuels through refinery processes. On the other hand, sunlight is a natural cheapest and clean energy source which can be used to produce hydrogen gas fuel. Therefore, photocatalytic hydrogen production through water splitting reaction has been considered as an excellent method to produce clean hydrogen utilizing earth-abundant sunlight [3].

2.1.2 Basic Principles of Photocatalytic Water Splitting

The direct water splitting into hydrogen and oxygen under solar light irradiation is the final objective of a photocatalytic hydrogen production system. Under ambient conditions, the splitting water is thermodynamically not favorable reaction having net Gibbs free energy change (ΔG^0) of 237 kJ/mol [4]. However, when semiconductor photocatalyst is irradiated with light in water, then thermodynamic equilibrium can be moved toward the required right direction for H_2 and O_2 production. Moreover, the overall splitting of water and full water splitting reaction is explained in Eqs. (2.1–2.4) [5].

M. B. Tahir and K. N. Riaz, *Nanomaterials and Photocatalysis in Chemistry*,
Materials Horizons: From Nature to Nanomaterials,
https://doi.org/10.1007/978-981-16-0646-5_2

The overall water splitting reaction:

$$H_2O + h^+ \rightarrow \frac{1}{2}O_2 + 2H^+ \tag{2.1}$$

$$2H^+ + 2e^- \rightarrow \frac{1}{2}O_2 + H_2 \tag{2.2}$$

The full water splitting reaction:

$$2H_2O \rightarrow O_2 + 4H^+ \tag{2.3}$$

$$4H^+ + 4e^- \rightarrow O_2 + 2H_2 \tag{2.4}$$

In general, the water splitting reaction through semiconductor photocatalysis consists of three main steps as shown in Fig. 2.1. The first step involves light absorption by semiconductor photocatalyst to create charge carriers (electrons and holes). In this process, when semiconductor is illuminated with incident light energy higher than its band gap energy (Eg), then electrons are excited from valence band (VB) of

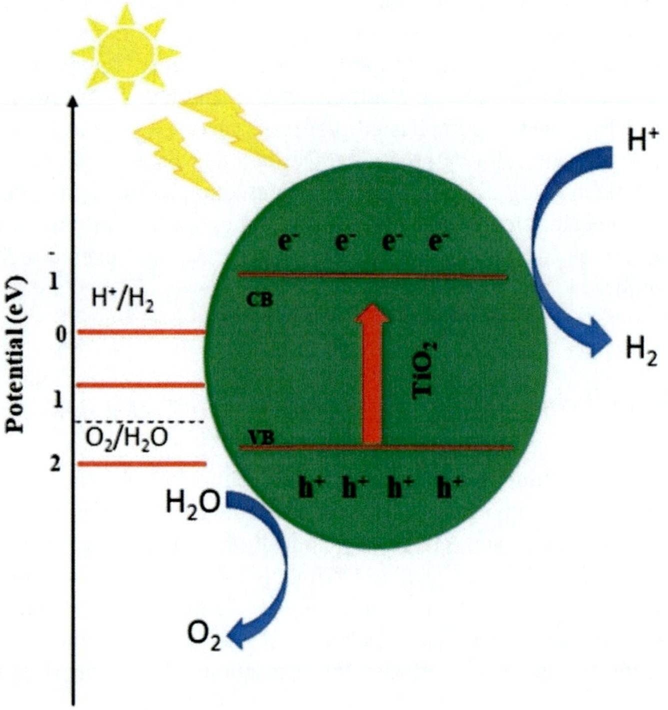

Fig. 2.1 Photocatalytic system for water splitting

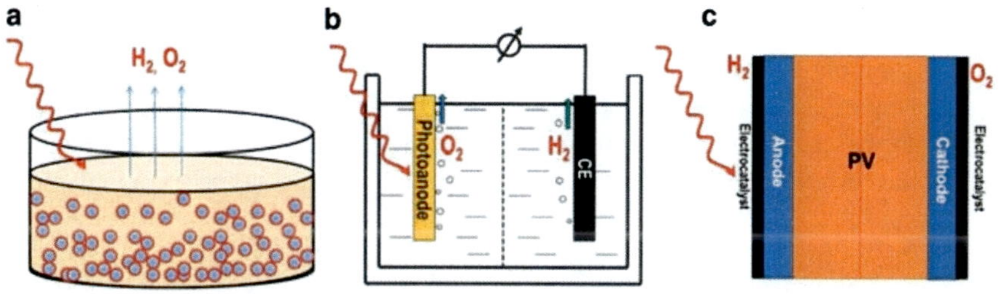

Fig. 2.2 Production of solar hydrogen via water splitting. **a** Particulate photocatalytic (PC) water splitting system, **b** photoelectrochemical (PEC) water splitting system, and **c** photovoltaic-photoelectrochemical (PV-PEC) hybrid system. Reproduced with permission from Ref. [4], Copy right (2017), Elsevier

material to the conduction band (CB), leaving behind holes in the VB. It is necessary that energy of incident light should be greater than the band gap energy of semiconductor photocatalyst for higher efficiency and effectiveness of the process.

The second stage in photocatalysis involves the separation and transportation of the photocreated charge carriers to the surface of semiconductor photocatalyst. This stage is critical because it determines that how much charge carriers generated in the first stage have participated in photocatalytic water splitting reactions. The recombination (bulk and surface) of the photo-excited charge carriers will inevitably happen during this stage, consuming the charge carriers in the form of light or thermal energy, which ultimately reduces the availability of the charges for water splitting. The last step is the redox reactions at the catalyst surface. The electrochemical process, in which the electron potential has to be more negative than 0 V versus normal hydrogen electrode (NHE) for reducing H^+ to H_2, and the hole potential should be more positive compared to oxidative potential of H_2O/O_2 (1.23 V vs. NHE, at pH = 0) [4–7]. The potentials of charge carriers (electrons and holes) originate from the energy levels of the semiconductor's CB and VB, which thus must be enough to fulfill the requirement of photocatalytic water splitting.

The photocatalytic water splitting mainly can be divided into three types, which are particulate photocatalyst (PC) system, photoelectrochemical (PEC) system, and photovoltaic-photoelectrochemical (PV-PEC) hybrid system (Fig. 2.2).

2.1.3 *Photocatalytic Water Splitting Types*

2.1.3.1 Particulate Photocatalytic Water Splitting

The particulate photocatalytic system is simple, low cost, and potential scalable photocatalytic water splitting in which photocatalyst powder is used in water and light is irradiated to split water for H_2 and O_2 evolution (Fig. 2.2a). However, the main disadvantage of this system is the separation of H_2/O_2 gas separation on large-scale during photocatalysis. Previously, microporous membranes based on molecular sieving effect were used to safely separate the both gases. For example, zeolite membrane has achieved great attention in gas separation and some reports have highlighted the current advancement in basic science and practical industrial applications [8–10].

2.1.3.2 Photoelectrochemical Water Splitting

In photoelectrochemical water splitting, firstly, electrode is developed using photocatalyst material and then a lower bias voltage is applied across the electrode to drive photocatalytic water splitting reaction. For complete formation of PEC working cell (Fig. 2.2b), at least one or both electrodes should be made from photosensitive semiconductor material having a space-charge layer developed at the interface of semiconductor/liquid. When light is illuminated, the photocreated charge carriers are separated by space charge field and transferred to semiconductor/liquid junction for chemical reaction [11]. Interestingly, separation system for H_2/O_2 mixture is not required in this PEC water splitting system because H_2 and O_2 are separately collected at both electrodes.

2.1.3.3 Photovoltaic-Photoelectrochemical Water Splitting

In PV-PEC hybrid system (Fig. 2.2c), the photocatalytic water splitting system for H_2 evolution composed of effective photovoltaic solar cell and water electrolysis. The PV-PEC system for water splitting has many advantages than PEC system without cost consideration. For example, the PEC system has some major obstacles such as lack of efficient light absorber, the corrosion of the semiconductor, and mismatch of band edge potentials of semiconductors for H_2 and O_2 production reactions. However, these mentioned problems are not present in PV-PEC water splitting system [4, 12, 13].

2.1.4 Performance Evaluation of Photocatalytic Water Splitting

In order to investigate the performance of a photocatalyst material in photocatalytic water splitting, the time-dependent evolution rate (μ mol h^{-1} g^{-1} catalyst) is the main parameter to be considered. Furthermore, the standard conditions including light source, reactor vessels, filters, and temperatures are also required to compare the efficiency of material as photocatalyst. The performance is usually dependent on the portion of solar energy utilized during reaction and is measured in the form of quantum yield (QY), apparent quantum yield (AQY), incident photon-to-current efficiency (IPCE), and absorbed photon-to-current efficiency (APCE). The hydrogen production rate can be evaluated according to Eq. (2.5) [14]:

$$Q = \frac{A}{V_m \times m \times t} \quad (2.5)$$

where Q denotes the actual amount of hydrogen gas produced in reaction, A represents the hydrogen peak area, V_m is for standard molar volume (22.4 L mol^{-1}), m is catalyst mass, and t represents time.

The following equations define the overall quantum yield and apparent quantum yield;

$$\text{Overall quantum yield } (\%) = \frac{\text{No. of reacted electrons}}{\text{No. of adsorbed photons}} \times 100\% \quad (2.6)$$

$$\text{Apparent quantum yield } (\%) = \frac{\text{No. of reacted electrons}}{\text{No. of incident photons}} \times 100\% \quad (2.7)$$

$$\text{Apparent quantum yield of } H_2 (\%) = \frac{2 \times \text{No.of evolved } H_2 \text{molecules}}{\text{No. of incident photons}} \times 100\% \quad (2.8)$$

$$\text{Apparent quantum yield of } O_2 (\%) = \frac{4 \times \text{No.of evolved } O_2 \text{molecules}}{\text{No. of incident photons}} \times 100\% \quad (2.9)$$

The solar energy to chemical energy conversion efficiency can be defined as STM, which represents the product of hydrogen evolution rate and Gibbs free energy via the following Eqs. (2.10);

$$STH = \frac{\text{Output energy as } H_2}{\text{Energy of incident solar light}} = \frac{r_{H_2} \times \Delta G}{P_{\text{Sun}} \times S} \times 100\% \quad (2.10)$$

2.2 Photocatalytic CO_2 Conversion

2.2.1 Importance of Photocatalytic CO_2 Conversion

The carbon dioxide (CO_2) is a cheapest and abundantly available resource and its conversion into useful hydrocarbons or fuel is a great challenge and an important research area [15]. Therefore, a lot of research was done to convert CO_2 into high energy products via complex chemical processes. Recently, photocatalytic CO_2

conversion into useful fuels or chemicals has gained tremendous attraction because of its ability to provide renewable resource of energy (fuels) and resolve current challenges and problems like global warming [16]. Therefore, the development of photocatalytic CO_2 conversion is the best technique because it gives us many valuable fuels including HCOOH, HCHO, CH_3OH, and CH_4.

2.2.2 *Basic Principles of Photocatalytic CO_2 Conversion*

It is well known that the natural photosynthesis occurs in green plants where plants convert CO_2 and water into O_2 and carbohydrates utilizing solar energy [15]. In some ways, the photocatalytic CO_2 conversion process is similar to photosynthesis, however, photocatalytic CO_2 conversion reaction uses semiconductor for photocatalysis as shown in Fig. 2.3. As was seen in photocatalytic water splitting reaction, the photocatalytic CO_2 conversion process also includes the same steps. Firstly, semiconductor absorbs light ($h\nu \geq E_g$) to generate electrons and holes. Secondly, the charge carriers are well-separated and transported to the photocatalyst surface. These photo-generated electrons and holes present in the conduction band and valence band, respectively, develop electron–hole pairs. Finally, on the photocatalyst surface, the redox reactions occur to convert CO_2 into fuels. It was also established that the photocatalytic conversion of CO_2 is an uphill reaction thermodynamically, where conduction band and valence band of the material should provide the necessary reduction and oxidation potential for CO_2 and H_2O, respectively. Generally, it is known that more negative conduction band edge compares to CO_2 reduction potential supports migration of photocreated electrons from conduction band to CO_2. Similarly, if valence band edge of semiconductor is more positive than oxidation potential of H_2O then easy transport of photo-excited holes occurs from valence band to water. Therefore, it can be said that photocatalytic reduction of CO_2 and oxidation of H_2O are simultaneously possible (Fig. 2.3). These reduction and oxidation reactions can

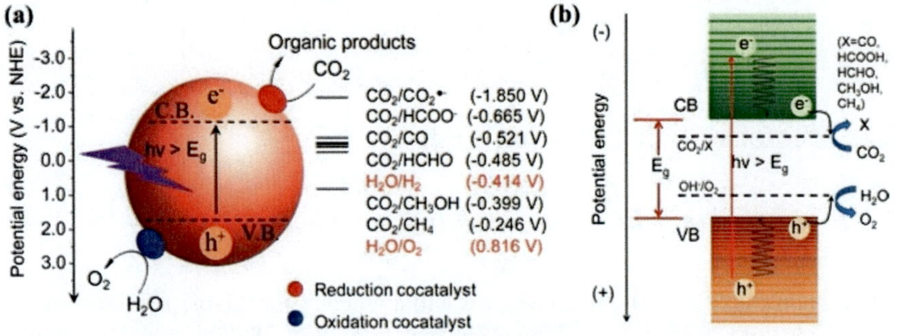

Fig. 2.3 **a** Photocatalytic mechanism representation and **b** relative energy levels of photocatalytic conversion of CO_2 on a semiconductor-based photocatalyst [16]

be well-understood through the following equations (Eqs. 2.11–2.18, at pH 7) [16].

$$CO_2 + e^- \rightarrow CO_2^{\cdot -} - 1.850\,V(\text{vs SHE}) \quad (2.11)$$

$$CO_{2(g)} + H_2O_{(l)} + 2e^- \rightarrow HCOO^-_{(aq)} + OH^-_{(aq)} - 0.665\,V(vsSHE) \quad (2.12)$$

$$CO_{2(g)} + H_2O_{(l)} + 2e^- \rightarrow CO_{(g)} + 2OH^-_{(aq)} - 0.521\,V(\text{vsSHE}) \quad (2.13)$$

$$CO_{2(g)} + 3H_2O_{(l)} + 4e^- \rightarrow HCHO_{(l)} + 4OH^-_{(aq)} - 0.485\,V(\text{vs SHE}) \quad (2.14)$$

$$CO_{2(g)} + 5H_2O_{(l)} + 6e^- \rightarrow CH_3OH + 6OH^-_{(aq)} - 0.399\,V(\text{vs SHE}) \quad (2.15)$$

$$CO_{2(g)} + 6H_2O_{(l)} + 8e^- \rightarrow CH_4 + 8OH^-_{(aq)} - 0.246\,V(\text{vs SHE}) \quad (2.16)$$

$$2H_2O_{(l)} + 2e^- \rightarrow H_{2(g)} + 2OH^-_{(aq)} - 0.414\,V(\text{vs SHE}) \quad (2.17)$$

$$2H_2O_{(l)} \rightarrow O_{2(g)} + 4H^+_{(aq)} + 4e^- + 0.816\,V(\text{vs SHE}) \quad (2.18)$$

Generally, the photocatalytic performance and the selection of CO_2 reduction by-product largely depend upon reaction system and photocatalyst material [6, 17]. The photocatalytic CO_2 reduction can yield a wide variety of products including CH_3OH and CH_4, HCOOH, CO, HCHO, CH_3CH_2OH, and elemental carbon [18]. The photocatalytic CO_2 reduction mainly is performed in gas or liquid phase system. It has been reported that CO and CH_4 are the main products in the gas state system. However, the main products in the liquid phase system include CH_3OH, CH_3COOH, and HCOOH [19–21]. The saturated aqueous solution of CO_2 is used in liquid phase reaction of photocatalysis for CO_2 conversion. However, in liquid phase system, the major and critical problem is lower solubility of CO_2 in aqueous solation which reduces its overall performance in photocatalysis [22]. Some researchers have enhanced the CO_2 solubility in water using NaOH, $NaHCO_3$, or Na_2CO_3 extra additives in the reactions. Secondly, using humidified CO_2, the gas phase photocatalytic reactions for CO_2 conversions can be performed.

2.2.3 *Performance Evaluation of Photocatalytic CO_2 Conversion*

The photocatalytic CO_2 conversion performance is normally evaluated by the following formula given below. Where R is mol h^{-1} g^{-1} of photocatalyst and its product can be taken in molar units (μ mol) or in units of concentration (ppm).

$$R = \frac{n(\text{product})}{\text{Time} \times m(\text{catalysts})} \quad (2.19)$$

The by-product of photocatalytic CO_2 reduction can be measured in apparent quantum yield (AQY) where efficiency of materials relies on amount of photocatalyst material used, intensity of light [23–27]. Therefore, the apparent quantum yield for CO_2 photo-conversion can be found through the following equations:

$$\text{Overall quantum yield } (\%) = \frac{\text{No. of reacted electrons}}{\text{No. of adsorbed photons}} \times 100\%. \quad (2.20)$$

$$\text{Apparent quantum yield}(\%) = \frac{\text{No. of reacted electrons}}{\text{No. of incident photons}} \times 100\%. \quad (2.2.1)$$

$$\text{Apparent quantum yield of CO}(\%) = \frac{2 \times \text{ No. of CO molecules}}{\text{No. of incident photons}} \times 100\% \quad (2.22)$$

$$\text{Apparent quantum yield of HCOOH}(\%) = \frac{2 \times \text{No. of HCOOH molecules}}{\text{No.of incident photons}} \times 100\% \quad (2.23)$$

$$\text{Apparent quantum yield of HCHO}(\%) = \frac{4 \times \text{No.of HCHO molecules}}{\text{No. of incident photons}} \times 100\% \quad (2.24)$$

$$\text{Apparent quantum yield of CH}_3\text{OH}(\%) = \frac{6 \times \text{No. of CH}_3\text{OH molecules}}{\text{No. of incident photons}} \times 100\% \quad (2.25)$$

$$\text{Apparent quantum yield of CH}_4(\%) = \frac{6 \times \text{No. of CH}_4\text{molecules}}{\text{No.of incident photons}} \times 100\% \quad (2.26)$$

2.3 Photocatalytic N_2 Fixation

2.3.1 Importance of Photocatalytic N_2 Fixation

The important role in sustaining the life on earth is played by natural nitrogen cycle because major biological molecules such as amino acids, proteins, and nucleic acids consist of nitrogen molecule [28, 29]. It is known that nitrogenize-based enzyme reduces nitrogen into ammonia (NH_3), proving major product for required for the biosynthesis of nitrogen-comprising molecules [30]. Similarly, ammonia itself is useful material to produce industrial high value added chemicals such as fertilizers [31]. Moreover, ammonia can be easily converted into liquid phase which support the easy transportation of high energy density fuels [32]. However, the synthesis of ammonia is accomplished via the Haber-Bosch process, which consumes a huge amount of energy and carriers a massive carbon footprint. It was estimated that Haber-Bosch process contributes the 1.5% of total CO_2 emissions (420 Mt per year) which is not favorable for efforts to overcome the issues like global warming. Therefore, photocatalytic nitrogen fixation offers a great opportunity for the development of green and advanced technologies to synthesize ammonia and is also a promising way toward more clean and cheap fuel product.

2.3.2 Basic Principles of Photocatalytic N_2 Fixation

The process of photocatalytic nitrogen fixation in principle resembles with water splitting as well as CO_2 conversion through photocatalysis [33]. This process of photocatalytic nitrogen fixation can be categories into several steps. Initially, the photoformed electrons are transported to conduction band of semiconductor leaving behind holes in valence band. In the next step, the photo-excited charge carriers migrate to the photocatalyst surface and participate in redox reactions to produce radicals. After multi-step reactions, the nitrogen is reduced into ammonia as shown in Fig. 2.4 and Eqs. (2.27–2.35) [34–37].

$$H_2O \rightarrow 1/2O_2 + 2H^+ + 2e^- 0.81\,V(\text{vs NHE atpH7}) \tag{2.27}$$

$$2H^+ + 2e^- \rightarrow H_2 - 0.42\,V(\text{vs NHEatpH7}) \tag{2.28}$$

$$N_2 + e^- \rightarrow N_2^- - 4.20\,V(\text{vs NHEat pH0}) \tag{2.29}$$

$$N_2 + H^+ + e^- \rightarrow N_2H - 3.20\,V(\text{vs NHEat pH0}) \tag{2.30}$$

$$N_2 + 2H^+ + 2e^- \rightarrow N_2H_2 - 1.10\,V(\text{vs RHE}) \tag{2.31}$$

$$N_2 + 4H^+ + 4e^- \rightarrow N_2H_4 - 0.36\,V(\text{vsNHEat pH0}) \tag{2.32}$$

$$N_2 + 5H^+ + 4e^- \rightarrow N_2H_5^{\mp} - 0.23\,V(\text{vs NHEat pH0}) \tag{2.33}$$

$$N_2 + 6H^+ + 6e^- \rightarrow 2NH_3 0.55\,V(\text{vs NHEat pH0}) \tag{2.34}$$

$$N_2 + 8H^+ + 8e^- \rightarrow 2NH_4^{\mp} 0.27\,V(\text{vs NHEat pH0}) \tag{2.35}$$

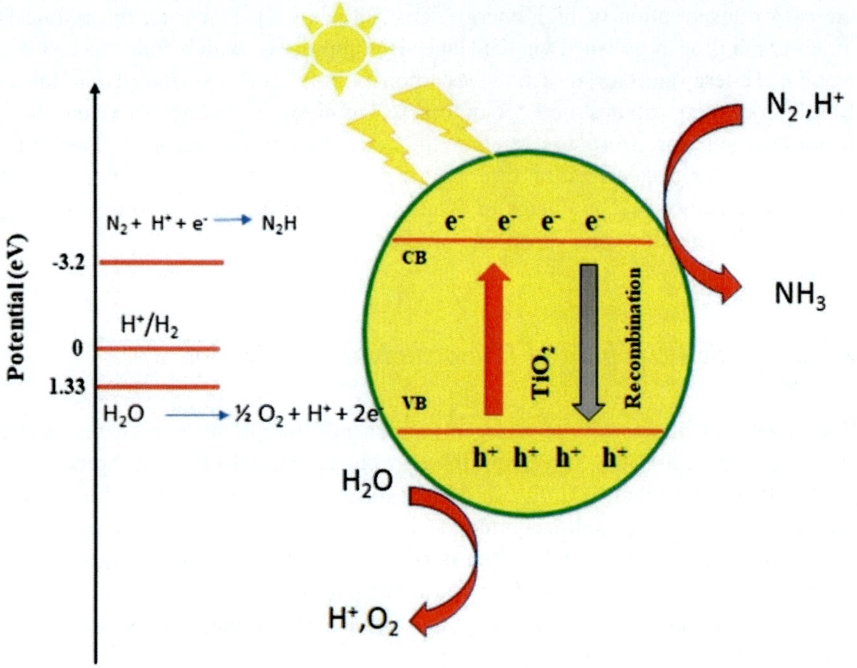

Fig. 2.4 Schematic diagram for the representation of semiconductor-based photocatalysis used for the reduction of N_2 to ammonia. The redox potentials (V vs. NHE at pH = 0) of water splitting and di-nitrogen hydrogenation are shown on the left side

In general, it is accepted that redox reactions are dependent on adsorbate reduction potential and energy band position of semiconductor materials [37–41]. For instance, it is necessary that the conduction band potential of material should be more negative than N_2 reduction potential while valence band edge should be more positive than the oxidation potential of water. The first electron transfer state (−4.16 V vs. NHE) represents the maximum energy transition and proton-induced electron transfer (−3.2 V vs. NHE) reactions, representing overall kinetics of the reaction [33]. Therefore, the activation of N_2 molecules for NH_3 formation requires visible-light active semiconductors with lower band gap energies, also providing the required thermodynamic reduction potential of N_2 to NH_3.

2.3.3 *Performance Evaluation of Photocatalytic N_2 Fixation Performance Evaluation*

The quantum yield (QY) can be used for the evaluation of photocatalytic efficiency of material in ammonia synthesis as shown in Eq. (2.36) [42].

$$QY(\emptyset) = \frac{\text{No. of molecules decomposed}}{\text{No. of photons absorbed}} \tag{2.36}$$

The quantum yield can also be calculated from rate of reaction and absorbed light intensity (I_a) through below equation [43]:

$$\emptyset = \frac{\text{Rate}}{I_a} \tag{2.37}$$

Besides quantum yield, the apparent quantum yield (AQY) also demonstrate the photocatalytic efficiency of ammonia synthesis process and can be found using the following formula [44]:

$$\text{Apparent quantum yield (AQY)} = \frac{\text{Rate}}{I_0} \tag{2.38}$$

where I_0 is incident light intensity.

References

1. S. Cao, L. Piao, X. Chen, Emerging photocatalysts for hydrogen evolution. Trends Chem. (2019)
2. J. Li et al., Frontiers of water oxidation: The quest for true catalysts. Chem. Soc. Rev. **46**, 6124–6147 (2017)
3. N. Serpone et al., Why do hydrogen and oxygen yields from semiconductor-based photocatalyzed water splitting remain disappointingly low? Intrinsic and extrinsic factors impacting surface redox reactions. ACS Energy Lett. **1**, 931–948 (2016)
4. R. Li, Chin. J. Catal. **38**, 5–12 (2017)
5. N. Fajrina, M. Tahir, A critical review in strategies to improve photocatalytic water splitting towards hydrogen production. Int. J. Hydrogen energy **44**, 540–577 (2019)
6. X. Chen, S. Shen, L. Guo, S.S. Mao, Chem. Rev. **110**, 6503–6570 (2010)
7. J.H. Yang, D.E. Wang, H.X. Han, C. Li, Acc. Chem. Res. **46**, 1900–1909 (2013)
8. F. Gallucci, E. Fernandez, P. Corengia, M. van Sint Annaland, Chem. Eng. Sci. **92**, 40–66 (2013)
9. C. Feng, K. Khulbe, T. Matsuura, R. Farnood, A. Ismail, J. Membr. Sci. Res. **1**, 49–72 (2015)
10. N. Kosinov, J. Gascon, F. Kapteijn, E.J.M. Hensen, J. Membr. Sci. **499**, 65–79 (2016)
11. K. Sivula, F. Le Formal, M. Grätzel, Chemsuschem **4**, 432–449 (2011)
12. O. Khaselev, J.A. Turner, Science **280**, 425–427 (1998)
13. H. Gerischer, Top. Appl. Phys. **31**, 115–172 (1979)
14. F. Deng, J.-P. Zou, L.-N. Zhao, G. Zhou, X.-B Luo, S.-L. Luo, Nanomaterial-based photocatalytic hydrogen production, in *Nanomaterials for the Removal of Pollutants and Resource Reutilization, Micro and Nano Technologies*, (2019) pp. 59–82
15. Z. Fu, Q. Yang, Z. Liu, F. Chen, F. Yao, T. Xie, Y. Zhong, D. Wang, J. Li, X. Li, G. Zeng, Photocatalytic conversion of carbon dioxide: From products to design the catalysts. J. CO_2 Utilization **34**, 63–73 (2019)
16. S.R. Lingampalli, M.M. Ayyub, C.N.R. Rao, Recent progress in the photocatalytic reduction of carbon dioxide. ACS Omega **2**, 2740–2748 (2017)
17. T. Inoue, A. Fujishima, S. Konishi, K. Honda, Photoelectrocatalytic reduction of carbon dioxide in aqueous suspensions of semiconductor powders. Nature **277**, 637–638 (1979)

18. A. Corma, H. Garcia, Photocatalytic reduction of CO_2 for fuel production: possibilities and challenges. J. Catal. **308**, 168–175 (2013)
19. D.S. Lee, Y.W. Chen, Photocatalytic reduction of carbon dioxide with water on In VO_4 with NiO cocatalysts. J. CO_2 Util. **10**, 1–6 (2015)
20. Z.Y. Wang, H.C. Chou, J.C.S. Wu, D.P. Tsai, G. Mul, CO_2 photoreduction using NiO/$InTaO_4$ in optical-fiber reactor for renewable energy. Appl. Catal. A-Gen. **380**, 172–177 (2010)
21. D.O. Adekoya, M. Tahir, N.A.S. Amin, g-C_3N_4/(Cu/TiO_2) nanocomposite for enhanced photoreduction of CO_2 to CH_3OH and HCOOH under UV/visible light, J. CO_2 Util. **18**, 261–274 (2017)
22. B. Parkinson, ACS Energy Lett. **1**, 1057–1059 (2016)
23. J. Mao, K. Li, T. Peng, Recent advances in the photocatalytic CO_2 reduction over semiconductors. Catal. Sci. Technol. **3**, 2481–2498 (2013)
24. L. Yuan, Y.-J. Xu, Photocatalytic conversion of CO_2 into value-added and renewable fuels. Appl. Surf. Sci. **342**, 154–167 (2015)
25. A.J. Morris, G.J. Meyer, E. Fujita, Molecular approaches to the photocatalytic reduction of carbon dioxide for solar fuels. Acc. Chem. Res. **42**, 1983–1994 (2009)
26. H. Takeda, K. Koike, H. Inoue, O. Ishitani, Development of an efficient photocatalytic system for CO_2 reduction using rhenium (i) complexes based on mechanistic studies. J. Am. Chem. Soc. **130**, 2023–2031 (2008)
27. S. Nahar, M.F.M. Zain, A.A.H. Kadhum, H.A. Hasan, M.R. Hasan, Advances in photocatalytic CO_2 reduction with water: A review. Materials **10**, 629 (2017)
28. V. Rosca, M. Duca, M.T. de Groot, M.T.M. Koper, Nitrogen cycle electrocatalysis. Chem. Rev. **109**, 2209–2244 (2009)
29. D.E. Canfield, A.N. Glazer, P.G. Falkowski, The evolution and future of Earth's nitrogen cycle. Science **330**, 192–196 (2010)
30. R. Schlogl, Catalytic synthesis of ammonia-a "never-ending story"? Angew. Chem. Int. Ed. **42**, 2004–2008 (2003)
31. I. Coric, B.Q. Mercado, E. Bill, D.J. Vinyard, P.L. Holland, Binding of dinitrogen to an iron-sulfur-carbon site. Nature **526**, 96–99 (2015)
32. G.-F. Chen, X. Cao, S. Wu, X. Zeng, L.-X. Ding, M. Zhu, H. Wang, Ammonia electrosynthesis with high selectivity under ambient conditions via a Li + incorporation strategy. J. Am. Chem. Soc. **139**, 9771–9774 (2017)
33. S. Mukherjee, D.A. Cullen, S. Karakalos, K.X. Liu, H. Zhang, S. Zhao, H. Xu, K.L. More, G.F. Wang, G. Wu, Metal-organic framework- derived nitrogen-doped highly disordered carbon for electrochemical ammonia synthesis using N_2 and H_2O in alkaline electrolytes. Nano Energy **48**, 217–226 (2018)
34. X. Chen, N. Li, Z. Kong, W.-J. Ong, X. Zhao, Photocatalytic fixation of nitrogen to ammonia: state-of-the-art advancements and future prospects, Mater. Horiz. https://doi.org/10.1039/c7mh00557a
35. C.J. van der Ham, M.T. Koper, D.G. Hetterscheid, Chem. Soc. Rev. **43**(5183), 5191 (2014)
36. L.-J. Guo, Y.-J. Wang, T. He, Chem. Rec. **16**, 1918–1933 (2016)
37. S. Sun, X. Li, W. Wang, L. Zhang, X. Sun, Appl. Catal. B **200**, 323–329 (2017)
38. W.J. Ong, L.L. Tan, S.P. Chai, S.T. Yong, A.R. Mohamed, Chemsuschem **7**, 690–719 (2014)
39. W.-J. Ong, L.-L. Tan, Y.H. Ng, S.-T. Yong, S.-P. Chai, Chem. Rev. **116**, 7159–7329 (2016)
40. L.K. Putri, L.-L. Tan, W.-J. Ong, W.S. Chang, S.-P. Chai, Appl. Mater. Today **4**, 9–16 (2016)
41. A. Fuertes, Mater. Horiz. **2**, 453–461 (2015)
42. F.K. Kessler, Y. Zheng, D. Schwarz, C. Merschjann, W. Schnick, X. Wang, M.J. Bojdys, Nat. Rev. Mater. **2**, 17030 (2017)
43. M.-H. Vu, M. Sakar, T.-O. Do, Insights into the recent progress and advanced materials for photocatalytic nitrogen fixation for ammonia (NH_3) production. Catalysts **8**, 621 (2018). https://doi.org/10.3390/catal8120621

44. H. Kisch, D. Bahnemann, Best practice in photocatalysis: comparing rates or apparent quantum yields? J. Phys. Chem. Lett. **6**, 1907–1910 (2015)

Chapter 3
Fundamentals of Photocatalysis for Environmental Remediation

Check for updates

3.1 Photocatalysis for Air Purification

The photocatalytic process using semiconductor nanomaterials is considered to be the most advanced technologies for air purification. The pollutants present in the air are particularly harmful compared to soil and water pollutants. The air pollutants are of two types including particulate matters (PM) and gas-phase pollutants. Both types of pollutants have been identified to be the main sources of air pollution. For first kind of air pollutants (PM), the source control, physical treatments including electrostatic precipitation, wet scrubbing, and filtration were found highly effective strategies [1–4]. On the other hand, photocatalysis is very efficient technique for the removal of gaseous pollutants Therefore, in the proceeding sections, we will focus on basic photocatalysis mechanism and reactions of gaseous pollutants including variety of volatile organic compounds (VOCs), NO_x, SO_x, CO, ozone, and odors.

3.1.1 Photocatalytic Oxidation of VOCs

Recently, the photocatalytic oxidation (PCO) is an advanced and widely used technique in the remediation of VOCs. Volatile organic compounds have various hazardous effects on human health and environment. The most detected indoor and outdoor VOCs come from aromatics, aldehydes, and halocarbons. For the better evaluation of photocatalysts performance, formaldehyde, trichloroethylene (TCE), toluene, and benzene were generally used in photocatalytic degradation studies [5–8]. The photocatalytic oxidation for VOCs degradation is not simple process because it has to deal with variety of hydrocarbons which are volatile at room temperature or particular conditions [9]. The basic degradation mechanism for VOCs is shown in Fig. 3.1, where upon light illumination of appropriate wavelength ($\lambda \geq E_g$), electrons transfer from valence band of semiconductor (e.g., TiO_2) to its conduction band and

M. B. Tahir and K. N. Riaz, *Nanomaterials and Photocatalysis in Chemistry*,
Materials Horizons: From Nature to Nanomaterials,
https://doi.org/10.1007/978-981-16-0646-5_3

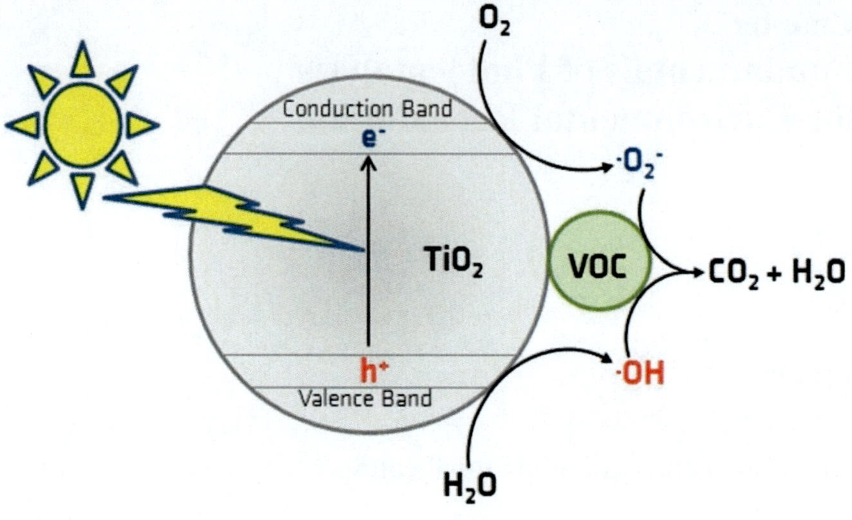

Fig. 3.1 Photocatalytic degradation mechanism of VOCs [6]

form an electron/hole pares via following reaction:

$$TiO_2 + h\vartheta \rightarrow h^+ + e^- \quad (3.1)$$

The photogenerated electrons in the presence of oxygen develop superoxide radical anion $O_2^{\bullet -}$. Simultaneously, the photo-excited holes can react with water to produce $OH^{\bullet}$ radicals. The formed radicals participate in mineralization of volatile organic substance into carbon dioxide and water as described by Eq. (3.2) [6, 9]:

$$O_2^{\bullet -} + OH^{\bullet} + VOCs \rightarrow CO_2 + H_2O \quad (3.2)$$

3.1.1.1 Photocatalytic Oxidation of Formaldehyde

Formaldehyde (HCHO) is an air pollutant which is produced from fossil fuel combustion. Both anthropogenic and natural sources are the primary source of formaldehyde because natural formaldehyde releases from wood, forest fire, and excretion of animals. Formaldehyde has been considered as the key indoor air pollutant whose exposure causes cancer and other serious diseases.

Peral et al. [10] first time investigated the photocatalytic oxidation of air pollutants in gas phase including formaldehyde and other air pollutants (acetone, 1-butanol and m-xylene). Then, plasmonic Au/TiO_2 photocatalyst was used by Zhu et al. [11] to degrade formaldehyde in air. In another report, Yang et al. [12] investigated the formaldehyde decomposition and proposed a degradation mechanism through the following equations (Eqs. 3.3–3.7).

$$HCHO + OH^{\bullet} \rightarrow CHO^{\bullet} + H_2O \tag{3.3}$$

$$CHO^{\bullet} + OH^{\bullet} \rightarrow HCOOH \tag{3.4}$$

$$CHO^{\bullet} + O_{2,ads}^{\bullet -} \rightarrow HCO_3^-(H^+) \rightarrow HCOOOH\ (+HCHO) \rightarrow HCOOH \tag{3.5}$$

$$\begin{aligned} &HCOOH(-H^+) \rightarrow HCOO^-(+OH^{\bullet}) \rightarrow H_2O + CO_2^{\bullet -}\text{or} \\ &HCOO^-(+h^+) \rightarrow H^+ + CO_2^{\bullet -} \end{aligned} \tag{3.6}$$

$$CO_2^{\bullet -} + (+[O] + OH^{\bullet} + h^+) \rightarrow CO_2 \tag{3.7}$$

3.1.1.2 Photocatalytic Oxidation of Toluene

The second air pollutant which was found in indoor as well as outdoor air is toluene. It can be said that toluene is the most studied model air pollutant in photocatalysis [13]. Bo et al. [14] observed the presence of alcoholic, aldehydic, and acidic intermediates from PCO of toluene. They found that h^+, $OH^{\bullet}$, and $O_2^{\bullet -}$ radicals formation on $CoCuMnO_x$ catalyst surface had vital role in the decomposition of toluene to above-mentioned products. The PCO of toluene at lower concentration was investigated by Quici et al. [15]. They studied the influence of catalyst film thickness of photocatalytic activity in the degradation of toluene. Recently, Nagendra et al. [16] used self-assembled air filters made of chitosan/AC/TiO_2 for the decomposition of toluene. This novel filter showed high remediation performance for toluene degradation (93%) and proposed a degradation mechanism as displayed in Fig. 3.2.

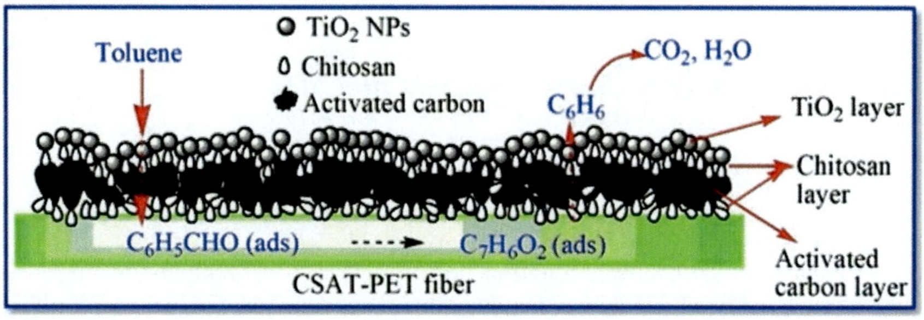

Fig. 3.2 Schematic representation of proposed remediation mechanism of self-assembled CSAT-PET. Reprinted with permission from Ref. [16], Copy right (2017), Elsevier

Blount et al. [17] used TiO_2 for the decomposition of gas-phase toluene and found a ring-like assembly with methyl groups.

3.1.1.3 Photocatalytic Oxidation of Benzene

The benzene is also toxic air pollutants and is derived from gasoline vapors, rubber, paint, plastic, cigarette smoke, and auto exhaust. Benzene has serious health threats even at lower exposure of it [18]. It is known that central precursor required for major industrial product is phenol, which is produced from the composition of benzene [19]. For instance, Han et al. [20] investigated photocatalysis for benzene conversion into phenol employing Co based catalyst. They observed higher photocatalytic activity when used $[Ru^{II}(Me_2phen)_3]^{2+}$ (Me_2phen = 4,7-dimethyl-1,10-phenanthroline) photocatalyst and $[Co^{III}(Cp^*)$ (bpy) $(H_2O)]^{2+}$ ($Cp^* = \eta^5$-pentamethylcyclopentadienyl and bpy = 2,2-bipyridine) as catalytic material due to the existence of $Sc(NO_3)_3$, O_2 (oxygen source), and electron source (H_2O). Mechanistic routes for benzene hydroxylation using visible photo-irradiation are shown in Fig. 3.3.

3.1.1.4 Photocatalytic Oxidation of Trichloroethylene (TCE)

Trichloroethylene (TCE) is also toxic air pollutant which is majorly utilized as solvent in chemical industry. The large-scale applications of TCE have polluted the industries and landfills. The TCE is chlorinated VOCs and typical air pollutant used for most photocatalytic studies [21]. The previous studies demonstrate that the developed intermediates of VOCs are more dangerous than other VOCs [22]. In a report, Ou et al. [23] studied the influence of humidity and oxygen on production of phosgene and DCAC. They found that optimized conditions by humidity and oxygen can lower the production amount of toxic intermediates and proposed a TCE degradation mechanism as shown in Fig. 3.4.

3.1.2 Photocatalytic Removal of NOx and SOx

3.1.2.1 Photocatalytic Removal of NO_x

NO_x are produced from human activities as well as natural processes occurring on our earth. For instance, volcanic action, transport from higher atmosphere toward lower environment, and few breakdown phenomenon of contaminants because of microbial function in ecosystem majorly produce nitric oxides as air pollutants [24]. Human activities are the major reason for NOx emissions in metropolitan areas including the combustion practices in stationary and transportable units (e.g., vehicles). The main

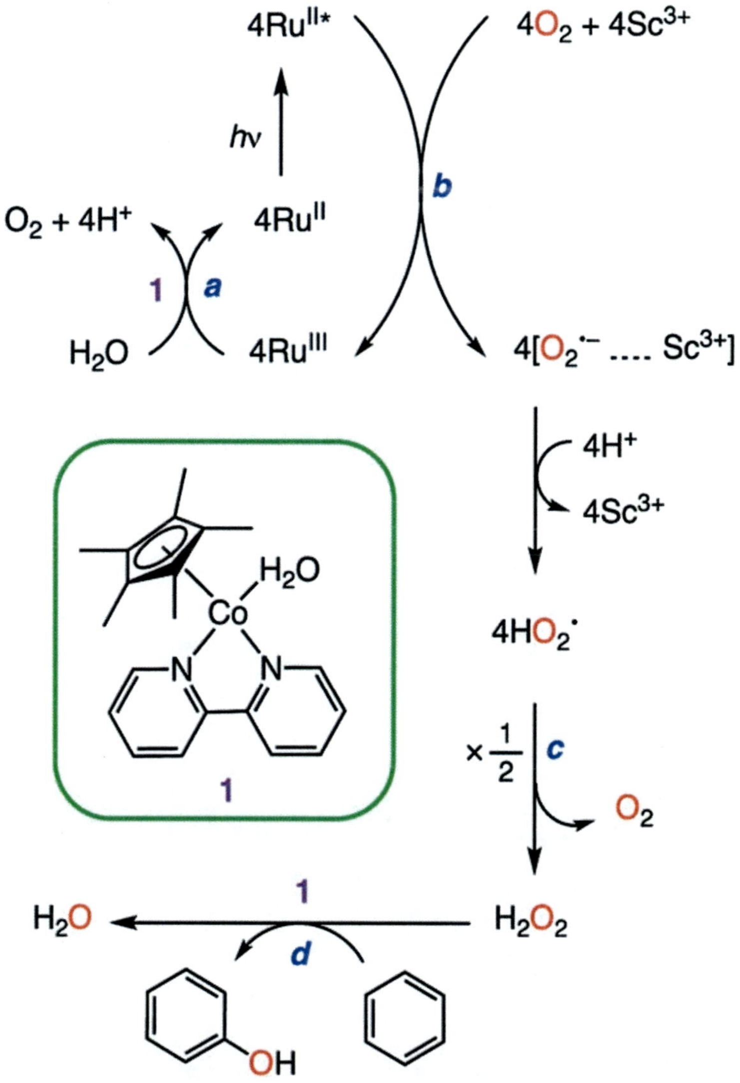

Fig. 3.3 Mechanistic diagram of the benzene photocatalytic hydroxylation [20]

forms of nitric oxides are NO and NO_2. The nitric oxides production has created environmental and health problems, such as global warming, tropospheric ozone production, acid rains, and people health sicknesses related to immune and respiratory organisms [25]. The most common methods for the remediation of NOx are those which emphasize on release control and conversion from NOx into N_2 through conversion process. The most favorable method for conversion of nitric oxides into useful products is photocatalysis which degrade NOx via three different steps including

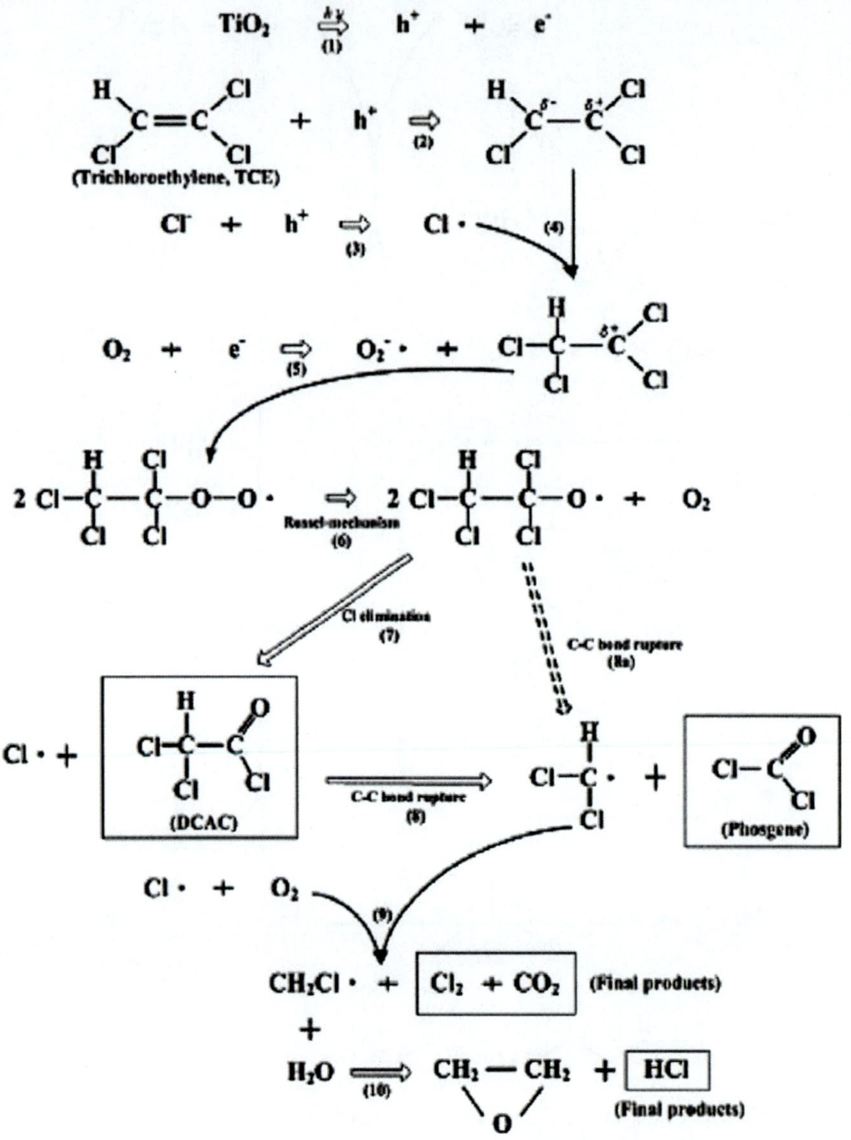

Fig. 3.4 Proposed mechanism of TCE photodegradation. Reprinted with permission from Ref. [23], Copy right (2007), Elsevier

photocatalytic decomposition, photocatalytic oxidation, and photocatalytic selective catalytic reduction process [26].

The photocatalytic decomposition of nitric oxide is to transfer NO into HNO_3 with the help of photocatalysis. The photocatalytic mechanism for conversion of NO into HNO_3 can be explained through the following equations [26].

$$NO + OH^{\bullet} \rightarrow HNO_2 \quad (3.8)$$

$$HNO_2 + OH^{\bullet}(ads) \rightarrow NO_2 + H_2O \quad (3.9)$$

$$NO_2 + OH^{\bullet}(ads) \rightarrow HNO_3 \quad (3.9)$$

$$NO + O^{\bullet -}_{2,ads} \rightarrow NO_3^- \quad (3.10)$$

$$2NO + O^{\bullet -}_{2,ads} + 3e^- \rightarrow 2NO_2 \quad (3.11)$$

$$3NO_2 + 2OH^- \rightarrow 2NO_3^- + NO + H_2O \quad (3.12)$$

Similarly, photocatalytic decomposition of nitric oxide to form N_2 follows other routes as described in previous reports [24, 27]. On the other hand, photocatalytic reduction process uses reducing agents including hydrocarbons, CO, or NH_3 to convert NO into N_2.

3.1.2.2 Photocatalytic Removal of SO_x

SOx in gas phase are considered to be one of the most hazardous air contaminants due to its serious effects on health and environment [24]. It also develops airborne particles which lead to degradation of visibility and climate change. It was found that higher closeness to SO_x contaminant air can cause respiration diseases [23]. Previous study describes that SO_2 could be degraded to sulfide and sulfate via developing highly reactionable oxygen species on TiO_2 thin-film-based photocatalyst [28].

$$SO_{2,ads} + O^{\bullet -}_{2,ads} \rightarrow SO^{2-}_{4,ads} \quad (3.13)$$

$$SO_{2,ads} + OH^{\bullet -}_{ads} \rightarrow HSO^-_{3,ads} \quad (3.14)$$

Xia et al. [29] have studied photocatalytic remediation of NO and SO_2 in gas phase under visible-light active $BiOI/Al_2O_3$ photocatalyst. The SO_2 conversion process and mechanism are displayed by these equations.

$$SO_2 + h^+ + 2H_2O \rightarrow H_2SO_4 + 2H^+ \quad (3.15)$$

$$SO_2 + OH^-_{ads} \rightarrow HSO_3 \quad (3.16)$$

$$SO_2 + O^{\bullet -}_{2,ads} \rightarrow SO_4^{2-} \quad (3.17)$$

3.1.3 *Photocatalytic Removal of Carbon Monoxide and Ozone*

CO is most common and poisonous air contaminant found in our earth [30]. It is produced from the burning of volatile organic compounds (hydro-carbons) and leads to fatal due to the presence of oxygen in blood. Generally, CO conversion to CO_2 is the final process which is accomplished in main degradation/oxidation reactions of volatile organic compounds. The carbon monoxide photocatalytic oxidation with oxygen can be seen in equations [31].

$$CO + h^+ \rightarrow CO^\bullet \tag{3.18}$$

$$O_2^{\bullet -} h^+ \rightarrow 2O \tag{3.19}$$

$$O^- + CO^\bullet \rightarrow CO_2 \tag{3.20}$$

The gas having blue color and pungent odor is present naturally thousand feet's above the earth's surface which is called ozone. This ozone layer has a positive role in the atmosphere due to absorbing ultraviolet radiations reaching the earth surface. Beside this advantage, it can become a toxic air pollutant when its amount is greater than 0.214 mg m^{-3} and causes hazardous effects on human health (e.g., headache, chest pain, eyes irritation, nose, and throat problems) [24]. Thus, photocatalytic process can be utilized for decontamination of ozone through the following equations given below [32].

$$O_3 + e^- \rightarrow O_3^- \tag{3.21}$$

$$OH^\bullet + O_3 \rightarrow O_4^{\bullet -} \tag{3.22}$$

$$O_3^{\bullet -} + O_3 \rightarrow O_4^{\bullet -} + O_2 \tag{3.23}$$

The ozone has high oxidization power, and therefore, in the atmosphere, it can be removed via oxidation of some VOCs present in air.

3.1.4 *Photocatalytic Deodorization*

Air purification via deodorization is an important process. The odor can be caused due to the S-containing compounds, bacteria, and hydrogen sulfide (H_2S). The basic mechanistic process of photocatalytic remediaition of S-containing organic pollutants is similar to that of volatile organic compounds [33]. However, the mechanism of photocatalytic decomposition of odorous bacteria is still not clear [34]. For an example, we will describe the photocatalytic removal of hydrogen sulfide to produce hydrogen gas as well as photocatalytic oxidation of H_2S into sulfate. Bhirud et al. [35] used N-TiO_2/graphene for the production of hydrogen through the following process.

$$H_2S + OH^{\bullet}_{ads} \rightarrow HS^{\bullet}_{ads} + H_2O \tag{3.24}$$

$$2HS^{-}_{ads} + 2h^{+} \rightarrow +2S + 2H^{+} \tag{3.25}$$

$$2H^{+}2e^{-} \rightarrow H_2 \tag{3.26}$$

Besides hydrogen evolution, photocatalytic oxidation processes for air purification can be also utilized to form sulfate from H_2S via these equations.

$$H_2S_{ads} + h^{+} \rightarrow H_2S^{+}_{ads} \rightarrow HS^{\bullet}_{ads} + H^{+} \tag{3.27}$$

$$OH^{\bullet}_{ads} + H_2S^{+}_{ads} \rightarrow HS^{\bullet}_{ads} + H_2O_{ads} \tag{3.28}$$

$$HS^{\bullet}_{ads} + O_{2,ads} \rightarrow HSOO^{\bullet}_{ads} \tag{3.29}$$

$$HSOO^{\bullet}_{ads} + O_{2,ads} \rightarrow SO_{2,ads} \tag{3.30}$$

3.2 Photocatalysis for Water Purification

Water pollution is the serious concern of todays world because of the industrial development and population increase. It is known that toxic pollutants have seriously contaminated surface as well as groundwater which is harmful for the whole environment. In developing countries, the supply of clean drinking water is a great challenge and task. Thus, there has been an increasing demand to provide efficient and advanced technologies to deal with most common hazardous water pollutants. In order to meet this challenge, advanced oxidation processes (AOPs) were widely used which depends on creation of highly reactive species. These reactive radicals can be produced via variety of routes including sonolysis, UV/H_2O_2, Fenton-like oxidation, electrochemical oxidation, photocatalysis, etc. [36]. Among these mentioned strategies, semiconductor photocatalysis is the highly advanced and effective method to decompose and remediate hazardous water pollutants due to its simplicity and high efficiency [37].

Table 3.1 Most common water pollutants used as model pollutant in photocatalysis

Pollutant types	Common examples
Contaminants of emerging concern (CECs)	• Antibiotics (amoxicillin, metronidazole) • Additives (polybrominated diphenyl ethers) • Dyes (methyl orange, rhodamine B, etc.) • Disinfectants (haloacetic acids, trihalomethanes) • Preservatives (dimethylphenols, parabens) • Pharmaceuticals (diclofenac, ibuprofen)
Endocrine disrupting compounds (EDCs)	• Alkylphenols • Bisphenol A • Heavy metals (Cr^{6+}, As^{5+}, Hg^{2+}, Cu^{2+}, Pb^{2+}) • Organotins (mono-butyltin, dibutylin, tributyltin) • Pesticides (atrazine, chlorpyrifos, diazinon) • Polycyclic aromatic hydrocarbon • Phthalates (di(2-ethylhexyl) phthalate, etc.) • Steroid hormones (17α-ethinylestradiol, 17β-estradiol, etc.)
Pathogenic germs	• Bacillus subtilis • Micrococcus lylae • Escherichia coli • Staphylococcus aureus • Salmonella typhi
Cyanotoxins	• Anatoxin-a • Microcystins • Nodularins • Cylindrospermopsin

This Table is reproduced from Refs. [37–39]

There are a large number of investigated water pollutants as given in Table 3.1, [37–39]. These pollutants can be categorized into groups based on their origin and influences on health and environment. The major water pollutants are contaminants of emerging concern (CECs), endocrine disrupting compounds (EDCs), pathogens, and cyanotoxins. CECs stand for any chemical present in water or in the environment which are not usually detected due to insignificant levels or whose adverse effects are not well known [40]. The EDCs are the compounds having various hazardous influences over the endocrine system of living species [41]. It is well recognized that the photocatalysis has good bactericidal and germicidal potential, which can easily breakdown the microbial cells via decomposition of cell wall and subsequent internal components destruction [42]. The cyanobacteria are photosynthetic bacteria mostly found in surface water systems, which is highly harmful for human health due to hazardous effects [39, 43].

3.2.1 Photocatalytic Degradation of CECs

Photocatalysis is highly effective and novel approach that has been tried for the decontamination of contaminants of emerging concerns (CECs). The CECs are natural and synthetic chemicals, and their transformation products are occurring in water systems throughout the world [44]. They consist of additives, disinfectants, antibiotics, pharmaceuticals, dyes, preservatives, etc., that are consistently being present in surface water, groundwater, drinking water, and municipal wastewater [45]. The CECs have a high potential of causing severe effects on ecosystem and human health [40]. The basic mechanism and principles for photocatalytic degradation of CECs (dye degradation here) using semiconductor photocatalyst (e.g., TiO_2) are briefly explained below [46, 47].

a. ***Photoexcitation.***

Photocatalytic process is started when a photogenerated electron is transferred from the valence band to the conduction band of TiO_2 due to irradiation of light photon having either higher or equal energy to band gap energy of photocatalyst material. The photo-excitation process also generates holes in the valence band. Finally, resulting into formation of (e^-/h^+) pair as displayed in Eq. (3.31).

$$\mathrm{TiO_2} + \mathrm{h}\vartheta \rightarrow \mathrm{TiO_2}\left(\mathrm{e}^-(\mathrm{CB}\right) + \mathrm{h}^+(\mathrm{VB})) \qquad (3.31)$$

b. ***Ionization of water.***

The photocreated holes in the valence band produce $OH^\bullet$ radicals by reacting with water molecules (3.32).

$$\mathrm{H_2O(ads)} + \mathrm{h}^+(\mathrm{VB}) \rightarrow \mathrm{OH}^\bullet(\mathrm{ads}) + \mathrm{H}^+(\mathrm{ads}) \qquad (3.32)$$

The $OH^\bullet$ radicals produced via ionization of water on the surface of TiO_2 are tremendously powerful oxidizing agent. These radicals attack adsorbed organic molecules on the photocatalyst surface, which causes mineralization of harmful organic contaminant.

c. ***Oxygen ionosorption.***

The photogenerated electrons (in conduction band) are captured via O_2 molecule to produce superoxide radical ($O_2^{\bullet -}$) as displayed in Eq. (3.33).

$$O_2 + e^-(CB) \rightarrow O_2^{\bullet -}(ads) \quad (3.33)$$

These superoxide radicals can not only participate in the oxidation practice but also overcome the recombination of photogenerated charge carriers, thus retaining neutralization of electron within TiO_2.

d. ***Protonation of superoxide.***

Superoxide radicals ($O_2^{\bullet -}$) developed through above-mentioned reactions create hydroperoxyl radical (HO_2) and then subsequently H_2O_2 which further degrades into reactive hydroxyl radicals ($OH^{\bullet}$) via following equations.

$$O_2^{\bullet -}(ads) + H^+ \leftrightarrow HOO^{\bullet}(ads) \quad (3.34)$$

$$HOO^{\bullet}(ads) \rightarrow H_2O_2(ads) + O_2 \quad (3.35)$$

$$H_2O_2(ads) \rightarrow OH^{\bullet}(ads) \quad (3.36)$$

$$\text{Dye} + OH^{\bullet} \rightarrow CO_2 + H_2O(\text{dye intermediates}) \quad (3.37)$$

$$\text{Dye} + h^+(VB) \rightarrow \text{oxidation products} \quad (3.38)$$

$$\text{Dye} + e^-(CB) \rightarrow \text{reduction products} \quad (3.39)$$

Both processes of oxidation and reduction generally take place on the surface of semiconductor photocatalyst. The complete dye degradation process is shown in Fig. 3.5.

3.2.2 Photocatalytic Reduction of EDCs

Endocrine disrupting compounds (EDCs) consist of big groups of emerging contaminants which are extensively dispersed in surface and groundwater systems. The EDCs can cause disruption in the hormonal system and create a negative health influence on neurological, reproductive, and immune systems [48–50]. In general, the main examples of EDCs are alkylphenols, bisphenol A, heavy metals, organotins, pesticides, polycyclic aromatic hydrocarbon, phthalate, and steroid hormones (Table 3.1). Different methods have been used to reduce or degrade these harmful pollutants; however, photocatalytic reduction technique is safe, effective, nontoxic,

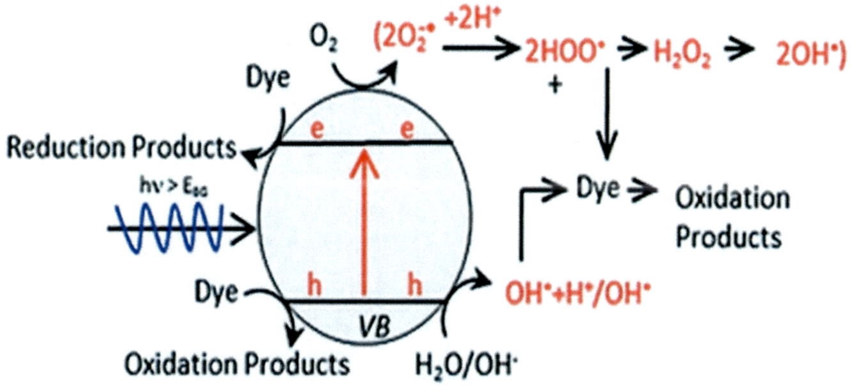

Fig. 3.5 Proposed reaction mechanism of dye degradation process. Reprinted with permission from Ref. [46], Copy right (2009), Elsevier

and economical for EDCs decontamination. Zhang et al. [50] found maximum reduction performance of heavy metal (Cd^{2+} and Pb^{2+}) ions over TiO_2/GO based photocatalysts. They proposed a photocatalytic reduction mechanism of TiO_2/GO nanocomposites as shown in Fig. 3.6, where Cd^{2+} and Pb^{2+} were easily reduced to Cd and Pb. The reduced as well as oxidized products and reactions involved in photocatalysis can be seen in the following equations.

$$TiO_2/GO + h\vartheta \rightarrow e^- + h^+ \tag{3.40}$$

$$e^- + O_2 \rightarrow O_2^{\bullet -} \tag{3.41}$$

$$2e^- + Cd^{2+} \rightarrow Cd \tag{3.42}$$

$$2e^- + Pb^{2+} \rightarrow Pb \qquad (3.42)$$

$$h^+ + H_2O \rightarrow H^+ + OH^- \tag{3.43}$$

$$h^+ + OH^- \rightarrow OH^\bullet \tag{3.44}$$

$$2OH^\bullet + Pb^{2+} \rightarrow PbO_2 + 2H^+ \tag{3.45}$$

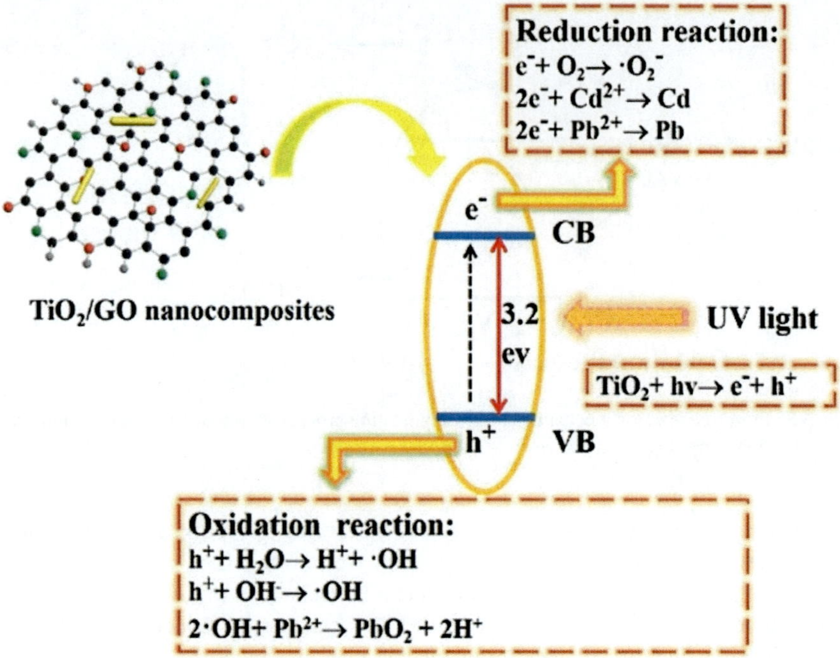

Fig. 3.6 Schematic representation of the charge transportation and separation in TiO_2-GO composites when illuminated with ultraviolet light and major routes for heavy metal ions reduction [50]

3.2.3 Photocatalytic Disinfection of Pathogenic Germs

Photocatalysis is unique process for using sunlight to drive the disinfection process of various pathogenic germs. When semiconducting material is irradiated with proper energy of light, reactive oxygen species (ROS) will be created to take part in disinfection of pathogenic microorganisms such as bacteria, viruses, spores, and protozoa [51]. The pathogenic germs treated via photocatalysis are bacillus subtilis, escherichia coli, micrococcus lylae, salmonella typhi, staphylococcus aureus, etc., as given in Table 3.1. The basic mechanism of photocatalysis oxidation process for inactivation of microorganisms is like disintegration process of contaminants because bacterial cells composed of 70–90% water and the remaining cellular contents of bacteria including polysaccharides, lipids, proteins, and nucleic acids are basically organic compounds which can be attacked by reactive oxygen species and consequently cause death of bacteria cell [52]. Figure 3.7 demonstrates the typical photocatalysis process of disinfection of microbial species [53]. The photocatalytic reaction is started via absorbing energy greater than band gap energy of semiconducting material. Then, the photocreated charges are separated and transported to surface of photocatalytic material, where they participate in range of redox reactions to generate ROSs. The photogenerated hydroxyl radicals ($OH^{\bullet}$), superoxide anions ($O_2^{\bullet-}$), singlet

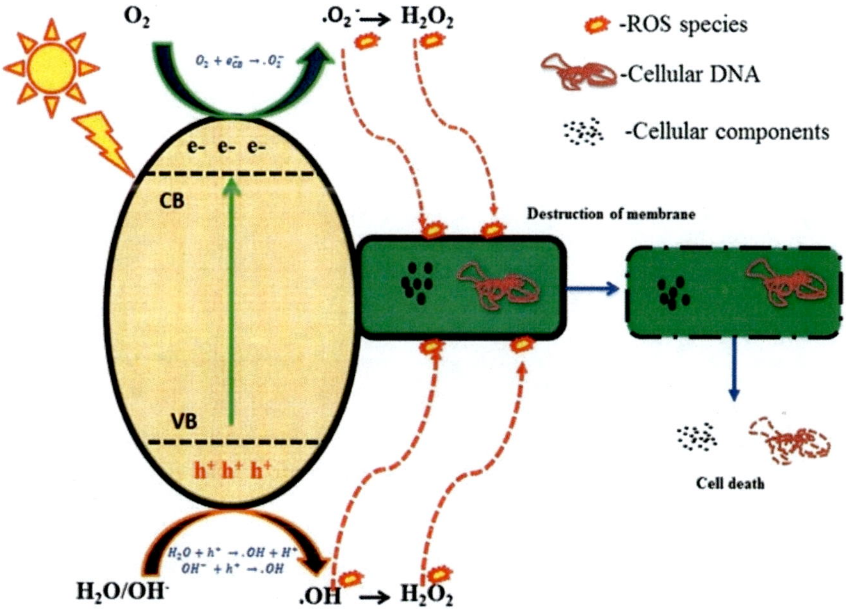

Fig. 3.7 Schematically illustrated mechanism of semiconductor-based photocatalysis disinfection of microbial species [53]

oxygen ($^1O_2$), and electrons (e^-) are the main ROSs responsible for the deactivation of microbial species [54–56].

3.2.4 Photocatalytic Disinfection of Cyanotoxins

Photocatalysis is gaining the popularity because of its higher efficiency, simplicity, and inexpensive nature in deactivation of water pollutants. The presence of cyanobacterial blooms (blue-green algae blooms) is a worldwide problem which influences the quality of water because of different cyanotoxins and other hazardous contaminants [57]. The cyanotoxins are a group of toxic secondary compounds produced from cyanobacteria inhabiting in freshwater and marine ecosystems. Moreover, the great problem is that some bloom-developing cyanobacteria yield hepatotoxic and neurotoxic toxins (microcystins from microcystic aeruginosa) [58]. These microcystins are highly toxic and harmful for the survival of several aquatic organisms, animals, and humans [59]. Microcystin-LR is commonly thought to be major extensive and toxic microcystin congener that causes to severe and chronic disease in animals [58, 60, 61]. Song et al. [61] have successfully removed M. aeruginosa and Microcystins-LR in an aqueous environment over g-C_3N_4/TiO_2 heterojunction photocatalyst under visible photo-irradiation. During the photocatalysis, the main reactive species were

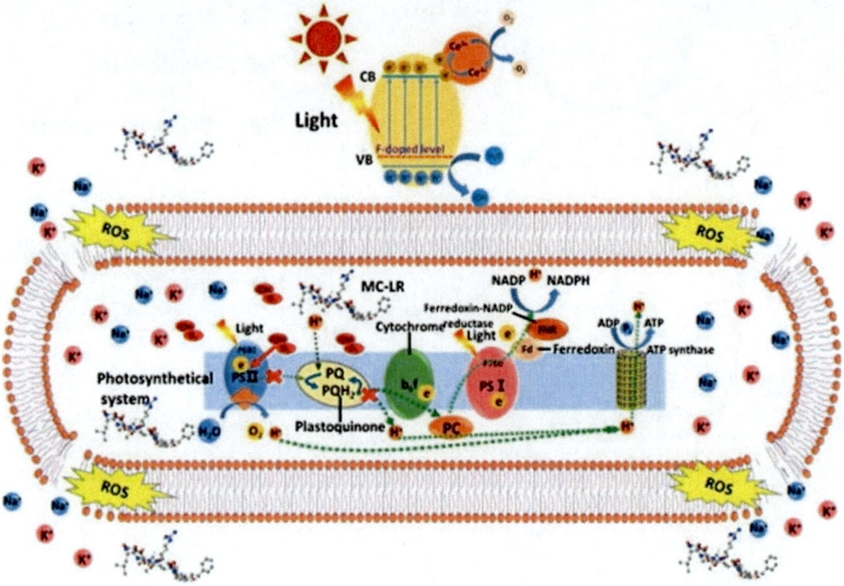

Fig. 3.8 Schematic diagram representing the possible reaction mechanism of photocatalytic method. Reproduced with permission from Ref. [58], Copy right (2017), Elsevier

found h^+ and $OH^•$ radicals responsible for deactivation of toxic compounds. Moreover, the photocatalytic process occurring on photocatalyst material surface not only disinfected the bacterial cells but also degraded debris.

Wang et al. [58] proposed a photocatalytic reaction scheme via employing F-Ce-TiO_2/EP floating catalyst (Fig. 3.8), where hydroxyl radicals ($OH^•$), superoxide anions ($O_2^{•-}$), and holes (h^+) were the primary reactive oxygen species (ROS) which played major role in deactivation of algal cells. They divided the whole photocatalytic disinfection process into three steps: (1) photocatalytic destruction of the cell walls and membranes by photogenerated ROS which can make irreversible damages to membrane protein, leading to electrolyte leaking, thus enhancing the solution conductivity [55, 62]; (2) photocatalysis enhanced photo-inhibition and pigment oxidation, where reactive oxygen radicals could outbreak pigments and active proteins, which reduce negative charge transport in W-W cycle, favoring the destruction of algal cells; (3) photocatalysis deactivation of metabolic products. These metabolic products are the secondary compounds released in dying procedure of algal cells. It was seen that these toxic products were easily adsorbed on the porous structure of floating photocatalyst and decomposed into water and carbon dioxide.

3.3 Parameters Affecting the Photocatalytic Degradation Efficiency

The effects of operating parameters including catalyst dosage, pollutant concentration, solution pH, inorganic salts, and oxidization species on photocatalytic efficiency were studied in the literature [63, 64]. These are the major factors which influence photocatalysis disintegration of water contaminants.

3.3.1 *Effect of Catalyst Dosage*

Catalyst dosage influences the efficiency of photocatalysis and varies with the type of pollutant. In general, the enhancement in photocatalytic material ratio can increase surface active sites on material and thus creates core reactive oxygen species for higher photocatalysis [62]. However, when catalyst concentration is increased above the optimum level, there will be definite adverse influences on performance of material because of agglomerated particles and lower opacity of solution, which ultimately results into lessening in radiation path length. Garg et al. [65] explored the impact of catalyst amount on activity of nitrogen and cobalt co-doped TiO_2 photocatalyst for mineralization of Bisphenol-A (BPA) when irradiated with light photons of visible portion. They found that the optimum concentration was 140 mg L^{-1} and increasing the catalyst above this level decreased the activity, which was ascribed to scattering of light because of dispersed particles.

3.3.2 *Effect of Pollutant Concentration*

The pollutant type and concentration is another major factor which affects the photocatalytic efficiency of material. Many investigators have studied the photocatalysis performance under same operating parameters, but varied the concentration of water contaminants [62, 66, 67]. They found that the percentage degradation was different for different concentration of pollutants. Moreover, it was observed that variation in dye concentration influences the intensity of light reaching the surface of photocatalytic material. Abdellah et al. [68] investigated effect of initial pollutant concentration on disintegration of methylene blue (MB) and varied concentration from 10–40 ppm. They noted a noteworthy reduction in photocatalysis action of dye degradation above optimum level of 10 ppm. They ascribed this decrease to photocatalyst surface coverage and screening effect of UV light due to enhanced dye adsorption at maximum loading. Similar outcome was found by Salama et al. [69] when they degraded methylene blue (MB) and indigo carmine (IC) dyes at different concentration of 10, 30, and 50 ppm. The maximum rate of degradation was achieved at the concentration of 10 ppm for both dyes.

3.3.3 *Effect of Solution PH*

Determination effect of solution pH on photocatalytic activity of material is highly tough assignment. It is well established that the degradation process of water pollutants can be completed through three possible reaction mechanisms including hydroxyl radical attacking, oxidizing process via hole reaction, and reducing process by conduction band negative charge [62]. pH is also considered as the major parameter that affects the adsorption practice because of its influence on surface charge of photocatalyst. For example, influence of solution pH (3–7) was investigated on photodisintegration process of methylene blue dye [68]. The authors observed an enhancement in photocatalysis efficiency when pH was increased from 3 to 7. They attributed this to MB dye structure type and amphoteric properties of photocatalytic material. More briefly, it was considered that TiO_2 surface in acidic aqueous environment having pH less than 5.8 gets a positive charge; however, when solution pH will be 6.8, it acquires a negative charge. At the pH values from 5.8 to 6.8, the surface of TiO_2 particles has no charge [70, 71]. Therefore, MB dye molecules were degraded more when TiO_2 particles has gain positive charge at pH higher than 6.8. Similar results were observed by Thu et al. [72], where basic pH at the surface of Cu/TiO_2 have a tendency to get negative charge, therefore, resulting into an increase adsorption and removal efficiency of MB due to growing electrostatic attraction between the positively charged pollutant and negative-charge photocatalytic material.

3.3.4 *Effect of Inorganic Salts*

Wastewater commonly includes elements including copper, iron, and phosphate, whose presence could decrease photocatalysis degradation efficiency. The key reason is that these elements might compete with contaminants during capturing surface active sites on photocatalytic material, leading to reduction in photocatalysis activity for targeted contaminants [47]. On the other hand, it was also observed that t zinc, calcium, and magnesium elements show less effect on photocatalysis action during contaminants disintegration process due to the reason that these elements exists in water environment with their highest oxidation states [73]. Wastewater of dyes industries comprises sufficient significant quantity of chlorides, carbonates, sulfates, and nitrates. These salts become the source of colloidal instability, and it was found that the CO_3^{2-} and HCO_3^- ions may scavenge the hydroxyl radicals ($OH^\bullet$), which decreases the photodegradation efficiency [46]. Similarly, decrease in photocatalysis activity due to chloride ions is because of hole (h^+) capturing characteristics of chloride ions [47]. Wang et al. [74] studied the influence of NO_3^- and SO_4^{2-} on the photocatalysis efficiency of reactive-red-2 dye and found an increase in dye removal due to the presence of these ions. In another report, similar outcomes were seen because NO_3^- ions have made quicker disintegration of azo dye when irradiated with light. This was ascribed to direct or indirect development of hydroxyl radicals [75].

3.3.5 *Effect of Oxidizing Species*

It is known that hydrogen peroxide (H_2O_2) and potassium peroxydisulfate ($K_2S_2O_8$) are powerful oxidizing species to increase photocatalysis oxidization for better photocatalytic efficiency [47, 76]. It was considered that hydrogen peroxide could display key part during in photocatalysis contaminant disintegration because it has the capability to harvest electrons from photocatalytic material's conduction band, thus decrease reunion of charges. However, hydrogen peroxide with concentration greater than critical may also act as hole or hydroxyl radical scavenger, which is detrimental to the photocatalytic activity [77–79]. Bizani et al. [80] found in their comparative study that H_2O_2 addition is highly influencing in photocatalysis of contaminants disintegration. In addition to this, the dye intermediates showed higher toxicity when $K_2S_2O_8$ was used as an oxidant than hydrogen peroxide, indicating lower toxicity removal efficiency.

References

1. J. Gao, X. Peng, G. Chen, J. Xu, G.L. Shi, Y.C. Zhang, Y.C. Feng, Insights into the chemical characterization and sources of PM2.5 in Beijing at a 1-h time resolution. Sci. Total Environ. **542**, 162–171 (2016)
2. Y. Zhuang, J.H. Pavlish, Fate of hazardous air pollutants in oxygen-fired coal combustion with different flue gas recycling. Environ. Sci. Technol. **46**, 4657–4665 (2012)
3. S.H. Byeon, B.K. Lee, B.R. Mohan, Removal of ammonia and particulate matter using a modified turbulent wet scrubbing system. Sep. Purif. Technol. **98**, 221–229 (2012)
4. C. Van der Heyden, P. Demeyer, E.I.P. Volcke, Mitigating emissions from pig and poultry housing facilities through air scrubbers and biofilters: State-of-the-art and perspectives. Biosys. Eng. **134**, 74–93 (2015)
5. P. Wolkoff, P.A. Clausen, B. Jensen, G.D. Nielsen, C.K. Wilkins, Are we measuring the relevant indoor pollutants? Indoor Air **7**, 92–106 (1997)
6. B. Mull, L. Möhlmann, O. Wilke, Photocatalytic degradation of toluene, butyl acetate and limonene under UV and visible light with titanium dioxide-graphene oxide as photocatalyst. Environments **4**, 9 (2017). 10.3390/environments4010009
7. A.A. Assadi, B. Abdelkrim, W. Dominique, Kinetic modeling of VOC photocatalytic degradation using a process at different reactor configurations and scalesm Int. J. Chem. Reactor Engin. 14 (2015)
8. J. Mo, Y. Zhang, Q. Xu, J.J. Lamson, R. Zhao, Photocatalytic purification of volatile organic compounds in indoor air: A literature review. Atmos. Environ. **43**, 2229–2246 (2009)
9. M. Pelaez, N.T. Nolan, S.C. Pillai, M.K. Seery, P. Falaras, A.G. Kontos, P.S.M. Dunlop, J.W.J. Hamilton, J.A. Byrne, K. O'Shea et al., A review on the visible light active titanium dioxide photocatalysts for environmental applications. Appl. Catal. B Environ. **125**, 331–349 (2012)
10. J. Peral, D.F. Ollis, Heterogeneous photocatalytic oxidation of gas-phase organics for air purification—acetone, 1-butanol, butyraldehyde, formaldehyde, and meta-xylene oxidation. J. Catal. **136**, 554–565 (1992)
11. X. Zhu, C. Jin, X-S. Li, J-L, Liu, Z-G. Sun, C. Shi, X. Li, A-M. Zhu, Photocatalytic formaldehyde oxidation over plasmonic Au/TiO_2 under visible light: Moisture indispensability and light enhancement. ACS Catal. **7**, 10, 6514–6524 (2017)
12. J.J. Yang, D.X. Li, Z.J. Zhang, Q.L. Li, H.Q. Wang, A study of the photocatalytic oxidation of formaldehyde on Pt/Fe_2O_3/TiO_2. J. Photochem. Photobiol. a-Chem. **137**, 197–202 (2000)

13. O. Debono, F. Thevenet, P. Gravejat, V. Hequet, C. Raillard, L. Lecoq, N. Locoge, Toluene photocatalytic oxidation at ppbv levels: Kinetic investigation and carbon balance determination. Appl. Catal. B Environ. **106**, 600–608 (2011)
14. L. Bo, S. Xie, H. Meng et al., Photocatalytic oxidation of gaseous toluene by visible-light-driven $CoCuMnO_x$: Performance and mechanism. Catal. Lett. **147**, 1623–1630 (2017). https://doi.org/10.1007/s10562-017-2058-9
15. N. Quici, M.L. Vera, H. Choi, G.L. Puma, D.D. Dionysiou, M.I. Litter, H. Destaillats, Effect of key parameters on the photocatalytic oxidation of toluene at low concentrations in air under 254 + 185 nm UV irradiation. Appl. Catal. B-Environ. **95**, 312–319 (2010)
16. V. Lekshmi Mohan, S.M. Shiva Nagendra, M.P. Maiya, Photocatalytic degradation of gaseous toluene using self-assembled air filter based on chitosan/activated carbon/TiO_2. J. Environ. Chem. Eng. **7**, 103455 (2019)
17. M.C. Blount, J.L. Falconer, Characterization of adsorbed species on TiO_2 after photocatalytic oxidation of toluene. J. Catal. **200**, 21–33 (2001)
18. J.H. Huang, X.C. Wang, Y.D. Hou, X.F. Chen, L. Wu, X.Z. Fu, Degradation of benzene over a zinc germanate photocatalyst under ambient conditions. Environ. Sci. Technol. **42**, 7387–7391 (2008)
19. R. Molinari, T. Poerio, Asia-Pac. J. Chem. Eng. **5**, 191 (2010)
20. J.W. Han, J. Jung, Y.-M. Li, W. Nam, S. Fukuzumi, Photocatalytic oxidation of benzene to phenol using dioxygen as an oxygen source and water as an electron source in the presence of a cobalt catalyst. Chem. Sci. **8**, 7119–7125 (2017)
21. V. Puddu, H. Choi, D.D. Dionysiou, G.L. Puma, TiO_2 photocatalyst for indoor air remediation: Influence of crystallinity, crystal phase, and UV radiation intensity on trichloroethylene degradation. Appl. Catal. B **94**, 211–218 (2010)
22. M. Mohseni, Gas phase trichloroethylene (TCE) photooxidation and byproduct formation: photolysis versus titania/silica based photocatalysis. Chemosphere **59**, 335–342 (2005)
23. H.H. Ou, S.L. Lo, Photocatalysis of gaseous trichloroethylene (TCE) over TiO_2: The effect of oxygen and relative humidity on the generation of dichloroacetyl chloride (DCAC) and phosgene. J. Hazard. Mater. **146**, 302–308 (2007)
24. Y. Boyjoo, H. Sun, J. Liu, V.K. Pareek, S. Wang, A review on photocatalysis for air treatment: From catalyst development to reactor design. Chem. Eng. J. (2016). https://doi.org/10.1016/j.cej.2016.06.090
25. J. Angelo, L. Andrade, L.M. Madeira, A. Mendes, An overview of photocatalysis phenomena applied to NOx abatement. J. Environ. Manage. **129**, 522–539 (2013)
26. J. Lasek, Y.H. Yu, J.C.S. Wu, Removal of NOx by photocatalytic processes. J. Photochem. Photobiol. C-Photochem. Rev. **14**, 29–52 (2013)
27. N. Bowering, G.S. Walker, P.G. Harrison, Photocatalytic decomposition and reduction reactions of nitric oxide over Degussa P25. Appl. Catal. B-Environ. **62**, 208–216 (2006)
28. Z. Topalian, G.A. Niklasson, C.G. Granqvist, L. Osterlund, Spectroscopic study of the photofixation of SO_2 on anatase TiO_2 thin films and their oleophobic properties. ACS Appl. Mater. Interfaces. **4**, 672–679 (2012)
29. D.H. Xia, L.L. Hu, C. He, W.Q. Pan, T.S. Yang, Y.C. Yang, D. Shu, Simultaneous photocatalytic elimination of gaseous NO and SO_2 in a $BiOI/Al_2O_3$-padded trickling scrubber under visible light. Chem. Eng. J. **279**, 929–938 (2015)
30. O. Rosseler, C. Ulhaq-Bouillet, A. Bonnefont, S. Pronkin, E. Savinova, A. Louvet, V. Keller, N. Keller, Structural and electronic effects in bimetallic PdPt nanoparticles on TiO2 for improved photocatalytic oxidation of CO in the presence of humidity. Appl. Catal. B-Environ. **166**, 381–392 (2015)
31. A. Nishimura, T. Hisada, M. Hirota, M. Kubota, E. Hu, Using TiO_2 photocatalyst with adsorbent to oxidize carbon monoxide in rich hydrogen. Catal. Today **158**, 296–304 (2010)
32. B. Ohtani, S.W. Zhang, S. Nishimoto, T. Kagiya, Catalytic and photocatalytic decomposition of ozone at room-temperature over titanium(iv) oxide. J. Chem. Soc.-Faraday Trans. **88**, 1049–1053 (1992)

33. H. Yao, A. Feilberg, Characterisation of photocatalytic degradation of odorous compounds associated with livestock facilities by means of PTRMS. Chem. Eng. J. **277**, 341–351 (2015)
34. S.B. Riffat, X.L. Ma, Performance testing of an optimal photocatalytic mop fan air cleaning system. Int. J. Energy Res. **37**, 1389–1396 (2013)
35. A.P. Bhirud, S.D. Sathaye, R.P. Waichal, J.D. Ambekar, C.J. Park, B.B. Kale, In-situ preparation of N-TiO_2/graphene nanocomposite and its enhanced photocatalytic hydrogen production by H2S splitting under solar light. Nanoscale **7**, 5023–5034 (2015)
36. A.R. Ribeiro, O.C. Nunes, M.F.R. Pereira, A.M.T. Silva, An overview on the advanced oxidation processes applied for the treatment of water pollutants defined in the recently launched Directive 2013/39/EU. Environ. Int. **75**, 33–51 (2015)
37. R. Fagan, D.E. McCormack, D.D. Dionysiou, S.C. Pillai, A review of solar and visible light active TiO_2 photocatalysis for treating bacteria, cyanotoxins and contaminants of emerging concern. Mater. Sci. Semicond. Process. **42**, 2–14 (2016)
38. V. Likodimos, D.D. Dionysiou, P. Falaras, Clean water: Water detoxification using innovative photocatalysts. Rev. Environ. Sci. Biotechnol. **9**, 87–94 (2010)
39. S. Merel, D. Walker, R. Chicana, S. Snyder, E. Baure‘s, O. Thomas, State of knowledge and concerns on cyanobacterial blooms and cyanotoxins. Environ. Int. **59**, 303–327 (2013)
40. J. Rivera-Utrilla, M. Sánchez-Polo, M.A. Ferro-García, G. Prados-Joya, R. Ocampo-Perez, Pharmaceuticals as emerging contaminants and their removal from water a review, Chemosphere **93**, 1268–1287 (2013)
41. M. Boas, U. Feldt-Rasmussen, K.M. Main, Thyroid effects of endocrine disrupting chemicals. Mol. Cell. Endocrinol. **355**, 240–248 (2012)
42. P.K.J. Robertson, J.M.C. Robertson, D.W. Bahnemann, Removal of microorganisms and their chemical metabolites from water using semiconductor photocatalysis. J. Hazard. Mater. **211–212**, 161–171 (2012)
43. A.A. De la Cruz, A. Hiskia, T. Kaloudis, N. Chernoff, D. Hill, M.G. Antoniou, X. He, K. Loftin, K. O'Shea, C. Zhao, M. Pelaez, C. Han, T.J. Lynch, D.D. Dionysiou, A review on cylindrospermopsin: The global occurrence, detection, toxicity and degradation of a potent cyanotoxins. Environ. Sci. Process Impacts **15**, 1979–2003 (2013)
44. G. Naidu, S. Jeong, Y. Choi, S. Vigneswaran, J. Membr. Sci. **524**, 565–575 (2016)
45. H. Luo, H. Li, Y. Lu, G. Liu, R. Zhang, Desalination **408**, 52–59 (2017)
46. M. Rauf, S.S. Ashraf, Chem. Eng. J. **151**, 10–18 (2009)
47. A. Ajmal, I. Majeed, R.N. Malik, H. Idriss, M.A. Nadeem, Principles and mechanisms of photocatalytic dye degradation on TiO_2 based photocatalysts: A comparative overview. RSC Adv. **4**, 37003 (2014)
48. J.O. Tijani, O.O. Fatoba, O.O. Babajide, L.F. Petrik, Environ. Chem. Lett. **14**, 27–49 (2016)
49. C. Jung, A. Son, N. Her, K.D. Zoh, J. Cho, Y. Yoon, J. Ind. Eng. Chem. **27**, 1–11 (2015)
50. H. Zhang, X. Wang, Na Li, J. Xia, Q. Meng, J. Ding, Lu Jie, Synthesis and characterization of TiO_2/graphene oxide nanocomposites for photoreduction of heavy metal ions in reverse osmosis concentrate. RSC Adv. **8**, 34241 (2018)
51. S. Das, S. Sinha, B. Das et al., Disinfection of multidrug resistant escherichia coli by solar-photocatalysis using Fe-doped ZnO nanoparticles. Sci. Rep. **7**, 104 (2017). https://doi.org/10.1038/s41598-017-00173-0
52. W. Wang, J.C. Yu, P.K. Wong, Photocatalysts for solar-induced water disinfection: New developments and opportunities. Mater. Sci. Forum **734**, 63–89 (2013)
53. P. Murugesan, J.A. Moses, C. Anandharamakrishnan, Photocatalytic disinfection efficiency of 2D structure graphitic carbon nitride-based nanocomposites: A review. J. Mater. Sci. **54**, 12206–12235 (2019)
54. S. Kim, K. Ghafoor, J. Lee, M. Feng, J. Hong, D. Lee, J. Park, Water Res. **47**, 4403–4411 (2013)
55. P.S.M. Dunlop, T.A. McMurray, J.W.J. Hamilton, J.A. Byrne, J. Photochem. Photobiol. A. **196**, 113–119 (2008)
56. W. Wang, Y.C. Yu, D. Xia, P.K. Wong, Y. Li, Environ. Sci. Technol. **47**, 8724–8732 (2013)

57. H. Xu, H. Lv, X. Liu, P. Wang, H.-L. Jiang, Electrolyte cations binding with extracellular polymeric substances enhanced Microcystis aggregation: Implication for Microcystis bloom formation in eutrophic freshwater lakes. Environ. Sci. Technol. **17**, 9034–9043 (2016)
58. X. Wang, X. Wang, J. Zhao, J. Song, J. Wang, R. Ma, J. Ma, Solar light-driven photocatalytic destruction of cyanobacteria by F-Ce-TiO_2/expanded perlite floating composites. Chem. Eng. J. **320**, 253–263 (2017)
59. A.C.Y. Yeung, P.M. D'Agostino, A. Poljak, J. McDonald, M.W. Bligh, T.D. Waite, B.A. Neilan, Physiological and proteomic responses of continuous cultures of Microcystis aeruginosa PCC 7806 to changes in iron bioavailability and growth rate. Appl. Environ. Microbiol. **19**, 5918–5929 (2016)
60. A.K. Makower, J.M. Schuurmans, D. Groth, Y. Zilliges, H.C. Matthijs, E. Dittmann, Transcriptomics-aided dissection of the intracellular and extracellular roles of microcystin in Microcystis aeruginosa PCC 7806. Appl. Environ. Microbiol. **81**, 544–554 (2015)
61. J. Song, X. Wang, J. Ma, X. Wang, J. Wang, S. Xia, J. Zhao, Removal of Microcystis aeruginosa and Microcystin-LR using a graphitic-C_3N_4/TiO_2 floating photocatalyst under visible light irradiation. Chem. Eng. J. (2018). https://doi.org/10.1016/j.cej.2018.04.182
62. K.M. Rezal, A.S.W. Kurny, F. Gulshan, Parameters affecting the photocatalytic degradation of dyes using TiO_2: a review. Appl. Water. Sci. **7**, 1569–1578 (2017)
63. B. Bayarri, J. Gimenez, D. Curco, S. Esplugas, Photocatalytic degradation of 2,4-dichlorophenol by TiO_2/UV: kinetics, actinometries and models. Catal. Today **101**, 227–236 (2005)
64. H. Zangeneh, A.A.L. Zinatizadeh, M. Habibi, M. Akia, M. Hasnain Isa, Photocatalytic oxidation of organicdyes and pollutants in wastewater using different modified titanium dioxides: a comparative review. J. Ind. Eng. Chem. **26**, 1–36 (2015)
65. A. Garg, T. Singhania, A. Singh, Photocatalytic degradation of bisphenol-a using N, Co codoped tio2 catalyst under solar light. Sci. Rep. **9**, 765 (2019). https://doi.org/10.1038/s41598-018-38358-w
66. E.M. Saggioro, A.S. Oliveira, T. Pavesi, C.G. Maia, L.F.V. Ferreira, J.C. Moreira, Use of titanium dioxide photocatalysis on the remediation of model textile wastewaters containing azo dyes. Molecules **16**, 10370–10386 (2011)
67. A. Giwa, P.O. Nkeonye, K.A. Bello, E.G. Kolawole, Solar photocatalytic degradation of reactive yellow 81 and reactive violet 1 in aqueous solution containing semiconductor oxides. Int. J. Appl. Sci. Technol. **2**, 90–105 (2012)
68. M.H. Abdellah, S.A. Nosier, A.H. El-Shazly, A.A. Mubarak, Photocatalytic decolorization of methylene blue using TiO_2/UV system enhanced by air sparging. Alexandria Eng. J. (2018) xxx, xxx–xxx
69. A. Salama, A. Mohamed, N.M. Aboamera et al., Photocatalytic degradation of organic dyes using composite nanofibers under UV irradiation. Appl. Nanosci. **8**, 155–161 (2018). https://doi.org/10.1007/s13204-018-0660-9
70. M. Salehi, H. Hashemipour, M. Mirzaee, Experimental study of influencing factors and kinetics in catalytic removal of methylene blue with TiO_2 nanopowder. Am. J. Environ. Eng. **2**, 1–7 (2012)
71. K.M. Joshi, V.S. Shrivastava, Removal of methylene blue dye aqueous solution using photocatalysis. Int. J. Nano Dimens. **2**, 241–252 (2012)
72. T.N.T. Thu, N.N. Thi, V.T. Quang, K.N. Hong, T.N. Minh, N. L.T. Hoai, Synthesis, characterisation, and effect of pH on degradation of dyes of copper-doped TiO_2. J. Exp. Nanosci. **11**, 3, 226–238 (2016). 10.1080/17458080.2015.1053541
73. M.N. Chong, B. Jin, C.W. Chow, C. Saint, Water Res. **44**, 2997–3027 (2010)
74. X. Wang, J. Jia, Y. Wang, J. Hazard. Mater. **185**, 315–321 (2011)
75. H. Zhu, J.-Y. Li, J.-C. Zhao, G.J. Churchman, Appl. Clay Sci. **28**, 79–88 (2005)
76. V. Augugliaro, C. Baiocchi, A. Bianco Prevot, E. García-López, V. Loddo, S. Malato, G. Marcí, L. Palmisano, M. Pazzi, E. Pramauro. Chemosphere, **49**, 1223–1230 (2002)
77. M.R. Dhananjeyan, E. Mielczarski, K.R. Thampi, P. Buffat, M. Bensimon, A. Kulik, J. Mielczarski, J. Kiwi, J. Phys. Chem. B **105**, 12046–12055 (2001)

78. A. Bozzi, M. Dhananjeyan, I. Guasaquillo, S. Parra, C. Pulgarin, C. Weins and J. Kiwi, J. Photochem. Photobiol. A **162**, 179–185 (2004)
79. S.S. Reddy, B. Kotaiah, Int. J. Environ. Sci. Technol. **2**, 245–251 (2005)
80. E. Bizani, K. Fytianos, I. Poulios, V. Tsiridis, J. Hazard. Mater. **136**, 85–94 (2006)

Chapter 4
Nanomaterials for Photocatalytic Energy Conversion

4.1 Nanomaterials for Photocatalytic Water Splitting

Nanomaterials have an important contribution in photocatalysis to split water into hydrogen and oxygen, where hydrogen (H_2) is a useful fuel for fuel cells to produce electricity. Moreover, nanomaterials structure, electronic and optical characteristics greatly control the performance of photocatalysis during each step. Most investigated photocatalytic nanomaterials include metal oxides, metal sulfides, metal-free-based nanomaterials, etc. [1]. The commonly used nanomaterials in photocatalytic H_2 production are reviewed in the following section.

4.1.1 Metal Oxide-Based Nanomaterials

There are several ways to classify the metal oxide-based nanomaterials such as metals which make the nanophotocatalyst, light absorption capability of nanomaterials (UV or visible-light active), and photocatalysis reaction type (H_2 production reaction, O_2 generation reaction, and/or overall splitting of water) [2]. In this section, first, we will highlight the progress on UV-light active (the materials which can absorb light less than 400 nm wavelength) nanomaterials including TiO_2, ZnO, and Ta_2O_5 used for photocatalysis to evolve H_2. In the next step, some visible-light responsive (those materials which have the capability to absorb light greater than 400 nm wavelength) photocatalysts such as WO_3, α-Fe_2O_3, and Cu_2O will be reviewed.

4.1.1.1 TiO_2

Titanium dioxide (TiO_2) is extensively reported nanomaterial in all photocatalytic application. It has a band gap of nearly 3.2 eV, where conduction band of TiO_2

M. B. Tahir and K. N. Riaz, *Nanomaterials and Photocatalysis in Chemistry*,
Materials Horizons: From Nature to Nanomaterials,
https://doi.org/10.1007/978-981-16-0646-5_4

is formed by Ti 3d orbitals and the valence band is developed by O 2p orbitals [2]. Due to its wide band gap, pristine TiO_2 can only response under ultraviolet light which is limited portion of solar spectrum (5%). Furthermore, for water splitting applications, TiO_2 conduction band potential is lightly less than water reduction potential and valence band potential is much greater than oxidation potential of water. Therefore, it is highly suitable nanomaterial in splitting of water application under ultraviolet photo-illumination. Variety of synthesis techniques were developed to fabricate TiO_2-based nanostructured with different morphologies such as one-dimensional TiO_2 (nanowires, nanorods, nanobelts, and nanotubes), two-dimensional TiO_2 (nanosheets and nanoplates), and three-dimensional TiO_2 hierarchical (nanobranches and nanoflowers) [3, 4]. The mainly used preparation methods are usually six including vapor deposition, electrospinning technique, hydrothermal, electrochemical anodiza-tion, sol-gel, and template [5–8]. These different morphologies of TiO_2 nanomaterials have greatly improved the efficiency of photocatalyst in splitting of water. Lai et al. [9] studied a material based on distribution of Pt nanoparticles on self-oriented nanotube arrays of TiO_2. The resulting nanostructured material demonstrated significantly improved PEC water splitting (higher production rate of H_2 up to 495 mmol h^{-1} cm^{-2}) when external bias of 0.3 V_{SCE} was used. Ji et al. prepared a hybrid consisting of 2D TiO_2 nanosheets/CdSe quan-tum dots which exhibited a ten-time higher photocatalytic hydrogen production activity compared with pure TiO_2 nanosheets [10]. Recently, Yang et al. prepared Au-TiO_2 nanofibers with an average diameter of ~160 nm via electrospinning combined followed by calcination treatment as shown in Fig. 4.1. The Au-TiO_2 nanofibers photocatalyst with 9% gold content has displayed higher efficiency in H_2 production (generation rate of H_2 is 12 440 μ mol/g h), exhibiting apparent quantum yield (APY) of 05.11% at 400 nm (Fig. 4.2). The observed APY for Au-TiO_2 nanofibers was much greater than P25 and bare TiO_2.

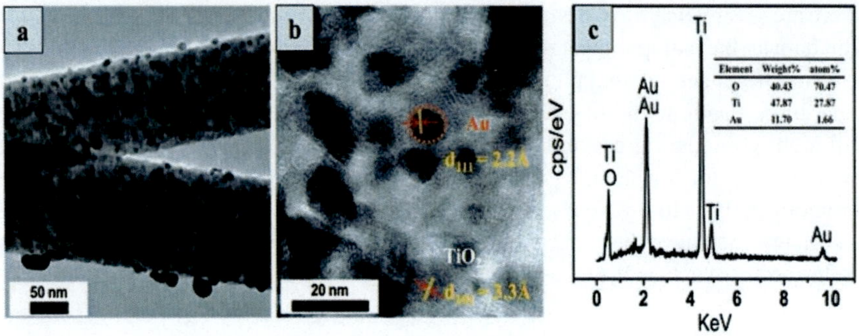

Fig. 4.1 **a** TEM image of TiO_2–Au-12 wt% composite based nanofibers (NFs); **b** HRTEM image of Au NPs embedded TiO_2 NFs representing the Au and TiO_2 lattice information; **c** EDS spectrum of TiO_2–Au-12 wt% composite sample [11]

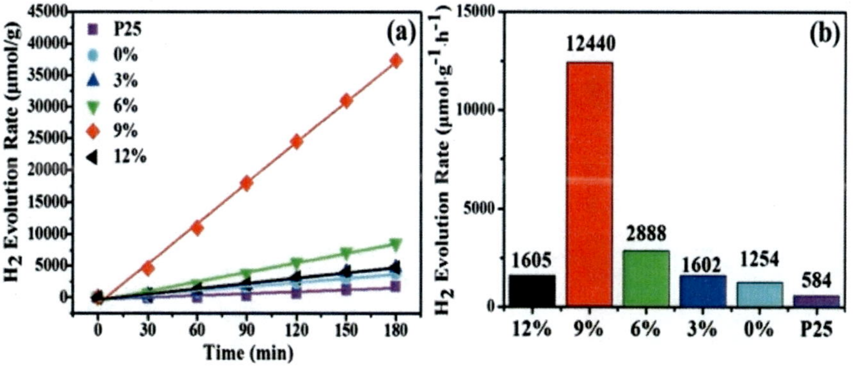

Fig. 4.2 Hydrogen production of P25, TiO_2 NFs, and TiO_2–Au composite nanofiber-based photocatalyst with various gold ratios under light illumination [11]

4.1.1.2 ZnO

Zinc oxide (ZnO) is highly chemically stable, nontoxic, and abundantly available semiconductor photocatalyst which showed superior performance in photocatalytic applications [12]. It has approximately similar band gap energy (3.37 eV) to TiO_2; therefore, it has attained tremendous attention and is considered to be the second highly efficient photocatalyst material after TiO_2. Compared to TiO_2, it has direct band gap and higher electronic properties along with thermal stability. However, its major drawback is lower optical absorption in visible portion of solar light due to wider band gap and is only active under ultraviolet light which share 5% of solar spectrum. Secondly, the photocreated electrons/holes recombine rapidly before reacting with molecules of water to produce hydrogen [13]. Therefore, in order to overcome both these problems, various methods were utilized to increase performance as photocatalyst in water splitting including morphology control, deposition of noble metal, sensitization of surface with dyes, composite formation with other semiconductor materials, and metal-nonmetal ions doping [14]. Therefore, researchers have studied N. C, and S-doped ZnO photocatalysts for higher visible-light absorption and photocatalytic activity. Bhirud et al. [15] have prepared nanostructured nitrogen-doped ZnO/graphene via simple in situ wet chemical technique. The nanostructured composite demonstrated a visible-light active improved performance in photocatalytic H_2 generation (~5070 μmol h^{-1}). They ascribed the excellent efficiency to the following reasons including high specific area, reduced band gap, and superior properties of graphene. Bai et al. synthesized 3DZnO/Cu_2O nanowires (NWs)-based photocatalytic nanomaterial having superior photoelectrocatalytic efficiency and a higher H_2 generation rate which was ~5.5 μmol h^{-1} cm^{-2} in only 20 min [16].

A major challenge for ZnO use in photoelectrochemical applications is chemically nonstability in water environment [17]. The chemical dissolution of ZnO occurs in strong acid and alkaline electrolyte-based solutions under dark conditions.

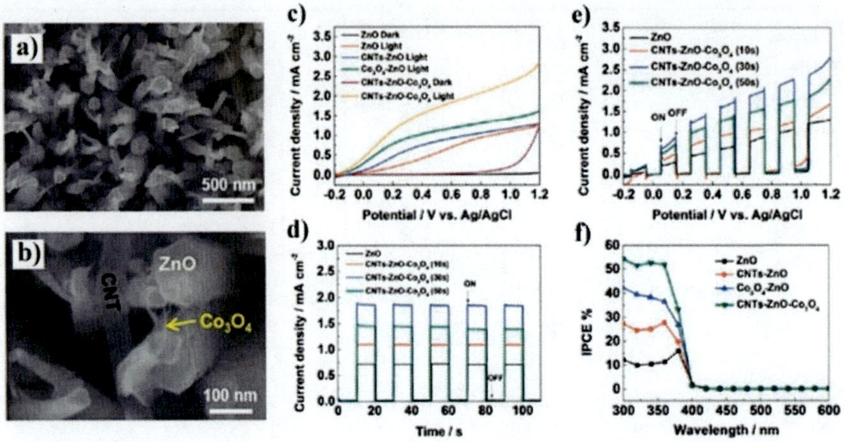

Fig. 4.3 **a** and **b** SEM images of Co_3O_4 coated ZnO nanowires (NWs) interlaced with carbon nanotubes (CNT), **c** current–voltage curves of pure ZnO NWs, binary CNT–ZnO (0.25 mg ml-1) NWs, Co_3O_4–ZnO (30 s) NWs and ternary CNT–ZnO–Co_3O_4 (30 s) NWs, **d** amperometric I–t curves of bare ZnO nanowires and ternary CNT–ZnO–Co_3O_4 nanowires (with different Co_3O_4 electro-deposition times) determined at 0.6 V versus Ag/AgCl, **e** current–voltage curves of pure ZnO NWs and ternary CNT–ZnO–Co_3O_4 NWs (with changing Co_3O_4 electro-deposition times) under chopped light illumination, and **f** IPCE spectra of pure ZnO NWs, CNT–ZnO NWs, Co_3O_4–ZnO NWs, and CNT–ZnO–Co_3O_4 NWs at 0.6 V versus Ag/AgCl. A 0.5 M Na_2SO_4 (pH 6.8) aqueous electrolyte was utilized for the photoelectrochemical properties. Reprinted from Ref. [18] with permission from RSC

The main strategies to avoid this challenge include modification of ZnO by applying over-coating on ZnO which can avoid direct link between interface of electrolyte and ZnO. Li et al. [18] investigated Co_3O_4 as catalytic layer over ZnO to enhance the stability of ZnO in aqueous environment. The nanostructures were interlaced with carbon nanotubes (CNTs) for the facilitation of charge carrier transportation and separation (Fig. 4.3a, b). Furthermore, Fig. 4.3c-f displays that cobalt oxide over-coating with an optimum thickness exhibited a photocatalytic role proved via cathodic shit of photocurrent onset-potential and also enhancement in photocurrent density. However, a thicker layer of Co_3O_4 had decreased the performance (Fig. 4.3d, e) due to lower absorption of light by ZnO.

4.1.1.3 Ta_2O_5

Tantalum oxide (Ta_2O_5) is a wider band gap (4 eV) n-type semiconductor which gained significant interest in photocatalysis field for hydrogen evolution because of good stable nature, and higher electronic properties in water splitting reactions [2]. Unfortunately, pure tantalum oxide was found inactive as photocatalyst for overall water splitting due to no response in visible portion of solar light. Many investigations focused on modification techniques like nanostructure formation, doping and/or composite formation which definitely improved efficiency of Ta_2O_5 as photocatalyst in water splitting reactions to produce hydrogen [19–21]. For example, mesoporous Ta_2O_5 after loading with NiO was observed to be superior in photocatalytic water splitting as compare to pristine tantalum oxide [22, 23]. There are other wide band gap semiconductors including Nb_2O_5, Ga_2O_3, ZrO_2, and their compounds which were also used in photocatalytic applications [24–29].

4.1.1.4 WO_3

Tungsten trioxide (WO_3) has excellent properties and is an n-type semiconducting nanomaterial which showed good photocatalytic activity for water splitting. Its band gap energy varies from 2.4–2.8 eV, dependent on its stoichiometric properties, phase assembly, and density of defects in WO_3. The main crystal phases of WO_3 are monoclinic, triclinic, tetragonal, hexagonal, cubic, and orthorhombic [2]. As compared to all other structures, the monoclinic structure has showed excellent efficiency in photocatalytic reactions [30]. WO_3 has good valence band potential to oxidize water for oxygen evolution. The hole-diffusion length (~150 nm) and mobility of electron (6.5 cm^2 V^{-1} s) of WO_3 are excellent. On the negative side, the reduction potential required for water is not provided by WO_3 because its conduction band potential is positive, which restrict its utilization in hydrogen production reactions via photocatalysis [31, 32]. Therefore, WO_3 was widely investigated for overall water splitting to produce oxygen via fabricating Z-scheme-based nanocomposites.

In a report, Zhang et al. studied a superior CdS/WO_3 composite material based on a Z-scheme system, which displayed visible-light responsive improved H_2 production because of lower probability of electron/hole pairs recombination [33]. Katsumata et al. [34] fabricated nanocomposite based on WO_3/g-C_3N_4 by simple heating process. The superior photocatalytic performance was obtained for 10 wt% WO_3/g-C_3N_4 sample with high hydrogen generation rate, which was two times higher compare to pristine g-C_3N_4 (Fig. 4.4a). Figure 4.4b represents that 10 wt%WO_3/g-C_3N_4 photocatalytic material displayed very stable nature for evolution of H_2 even after 30 h photocatalytic reaction time. Higher photocatalytic efficiency was ascribed to Z-scheme development system based on WO_3/g-C_3N_4 heterostructure, which have excellent charge separation capability for photo-excited electron–hole pairs (Fig. 4.4c). Liu et al. observed lower band gap energy (2.1 eV) and higher visible-light absorption for S/I-doped WO_3 photocatalytic material [35]. In another study, higher photo-catalytic efficiency of WO_3 nanoparticles modified by platinum with spherical macropores was observed via spray drying precursor process [36]. The authors observed that the photocatalyst prepared using PS/WO_3 mass ratio of 0.320 along with Pt co-catalyst exhibited higher photocatalysis performance compare to pristine WO_3 nanopar-ticles.

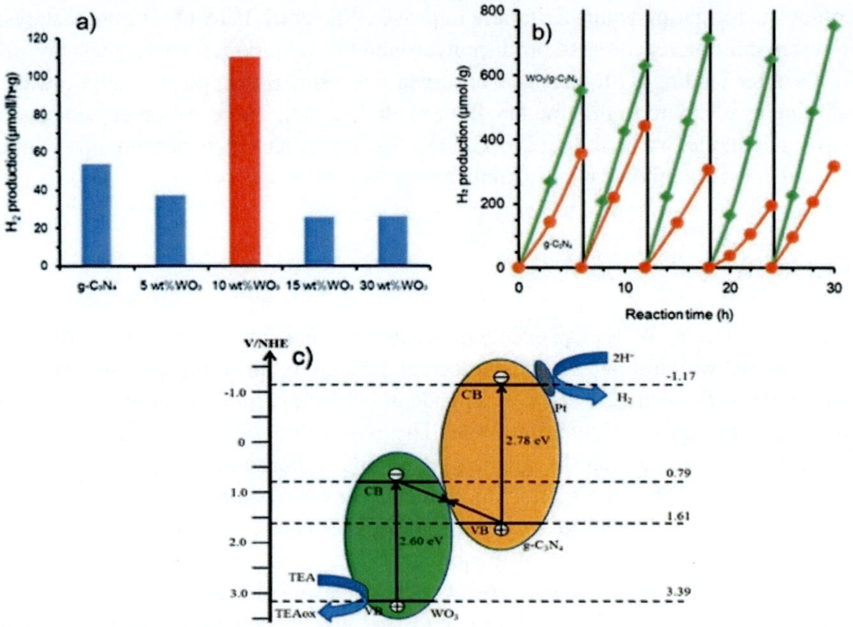

Fig. 4.4 **a** Photocatalytic hydrogen generation using g-C_3N_4 and WO_3/g-C_3N_4 nanocomposite under artificial solar light illumination, **b** time courses of H_2 generation by g-C3N4 and 10 wt% WO_3/g-C_3N_4 nanophotocatalyst using artificial solar light illumination, and **c** schematic representation of Z-scheme photocatalysis mechanism of WO_3/g-C_3N_4 nanocomposite photocatalyst [34]. Reprinted with permission from Ref. [34], Copyright (2014), RSC

4.1.1.5 α-Fe_2O_3

Hematite (α-Fe_2O_3) is a promising semiconducting material due to its lower band gap of 2.1 eV, better stability in oxidative environments, low cost, and easy availability [2]. Its more positivity of edge potential of conduction band edge compare to water reduction potential is not favorable for good photocatalytic activity. Thus, α-Fe_2O_3 in pure form is not good material to obtain higher photocatalysis efficiency for H_2 production. On the other hand, α-Fe_2O_3 valence band potential is higher (positive) than water oxidation potential and this property enables it to achieve higher performance photocatalytic water oxidation reaction [37, 38]. Unfortunately, α-Fe_2O_3 has some disadvantages including lower electrical conductivity, limited lifetime (~10 ps) of charge carriers, lower hole-diffusion length (~2–4 nm), which greatly limits its practical applications in PEC water oxidation [39]. However, in photoelectrochemical water splitting field, α-Fe_2O_3 has gained good intention and variety of strategies were developed including morphological variation, doping, heterostructure development, and modification of surface sensitization to increase the electronic, structural, as well as photocatalytic properties of hematite [40–43]. For example, the surface modification of α-Fe_2O_3 by thin deposition passivative overlayers of nanomaterials (metal oxide, metal hydroxide, graphene, etc.) was found highly positive and influencing for superior charge transport and separation at the interface of liquid-solid interface [44–46]. In a report, thin layer of TiO_2 (0.8 nm) was used for passivation of α-Fe_2O_3 photoanode by atomic layer deposition (ALD) technique [47]. This optimum thin TiO_2 layer exhibited enhanced photoelectrochemical efficiency of α-Fe_2O_3 due to well charge separation during photocatalysis.

4.1.1.6 Cu_2O

Copper oxide (Cu_2O) is a typical semiconductor nanomaterial (p-type) which has a direct band gap energy of about 2 eV. Copper oxide conduction band potential is more negative than water reduction potential and valence band just provide the required potential for oxidation of water. Therefore, due to suitable thermodynamics, it has the ability to drive both water reduction as well as oxidation half reactions. However, redox potentials of Cu_2O reduction and oxidation lie within its band gap, thus, it is somehow unstable in aqueous media during light irradiation [2]. Hara et al. first time used Cu_2O in photocatalysis to split water overall using light illumination of visible. They observed that the Cu_2O can sustain its performance for more than 1900 h during photocatalysis [48]. There are also some other reports on the use of Cu_2O for improved photocatalysis performance in H_2 production using sacrificial agents of electron accepters [49–51].

ALD is the most attractive technique to passivate the surface of semiconductor photocathodes [52]. The resulting Cu_2O photocathodes by ALD overlayers are good to increase transportation of photocreated negative charges to electrolyte which favors water reduction. At the same time, the sweeping of holes into bulk Cu_2O occurs. Moreover, in order to improve photocurrent and photostability of copper oxide, Kimura el al. utilized cathodic reduction with varying time intervals of electrochemical deposition (ED) to convert Cu_2O into CuO followed by depositing protecting n-type layers with heat treatment [53]. The superior photoresponse was found for Al_2O_3 drop-coated sample with 20 min of cathodic reduction among Al_2O_3, ZnO, and TiO_2 overlayers. The development of composite is another excellent strategy to enhance PEC efficiency of Cu_2O photocathode. Yang et al. synthesized Cu_2O-CuO bi-layered heterostructure by pulse chronoamperometric (r-DPPC) deposition [54]. The obtained bilayer composite was efficient than bare Cu_2O for PEC water splitting.

There are some other visible-light responsive metal oxides photocatalyst including Bi-based ($BiVO_4$, Bi_2WO_3) [55–57], Ag-based ($AgNbO_3$, Ag_3VO_4, $AgInW_2O_8$, $AgSbO_3$) [58, 59], lead-based ($PbWO_4$) [60], molybdenum-based ($PbMoO_4$) [61], Tin-based oxides ($SnNb_2O_6$) [62], etc., which were also reported in photocatalysis reactions to split water.

4.1.2 Metal Sulfide-Based Nanomaterials

Metal-based sulfides are good photocatalytic materials to split water for generation of H_2 gas. Therefore, some typical sulfide for photocatalysts water splitting is briefly reviewed below.

4.1.2.1 CdS

The cadmium sulfide having wurtzite crystal structure has attracted much attention compare to all other available metal sulfide photocatalytic semiconductors. Due to smaller band gap of 2.4 eV, CdS can efficiently utilize visible-light part of solar spectrum. Secondly, the conduction band potential of CdS is (~0.52 V vs. NHE) reasonably sufficient to reduce H_2O, and similarly, the valence band potential of CdS is also (1.5 V vs. NHE) appropriate for H_2O oxidation of [63]. However, the most common problem of metal sulfides photocatalysts including CdS is prone to photocorrosion. Therefore, many efforts have been taken to overcome this photocorrosion [64, 65]. The enhancement in photocatalysis efficiency and durability of CdS for hydrogen generation was obtained using CdS with different morphology, designing CdS-based hybrids with other semiconductor, polymers, or nanomaterials of high surface area.

Bao et al. [66] developed CdS with special hollow and sheet-like nanostructure possessing higher specific area. The high surface area has shortened the transport distance of electrons and holes from bulk to surface of photocatalyst, which enhanced the photocatalytic performance. Similarly, Sathish et al. [67] studied CdS nanoparticle with smaller grain size and higher surface area, showing the best photocatalysis efficiency for hydrogen production. The heterojunction formation has been proved to be the highly active strategy to achieve effective photocatalytic process to produce hydrogen because of enhanced separation ability of electron/hole. Yu et al. [68] prepared CdS-NiS heterostructure photocatalysts, which greatly improved the H_2 production efficiency. As displayed in Fig. 4.5a, about 20 nm particles of NiS were decorated on the surface of CdS nanorods homogeneously, which successfully formed a heterostructure between NiS and CdS.

This formation of p–n heterostructures has greatly supported the charge transfer between the two semiconductor photocatalyst, leading to lower charge carrier recombination (Fig. 4.5b, c). It can be seen that remaining positive holes in n-type photocatalyst were moved to p-type photocatalyst, giving a negative electron. The distribution of charge carriers continuous until a Fermi-level equilibrium is reached [69–71]. Therefore, charges reunion rate is effectively reduced due to cooperative influence between two semiconductors. Moreover, the photocatalysis hydrogen evolution rate for CdS-NiS nanocomposite with 5% nickel sulfide loading was found superior than pristine CdS (Fig. 4.5d). Further incorporation of NiS into CdS above optimum level (5 wt%) has reduced the photocatalytic performance of heterostructure due to less number of redox reaction sites.

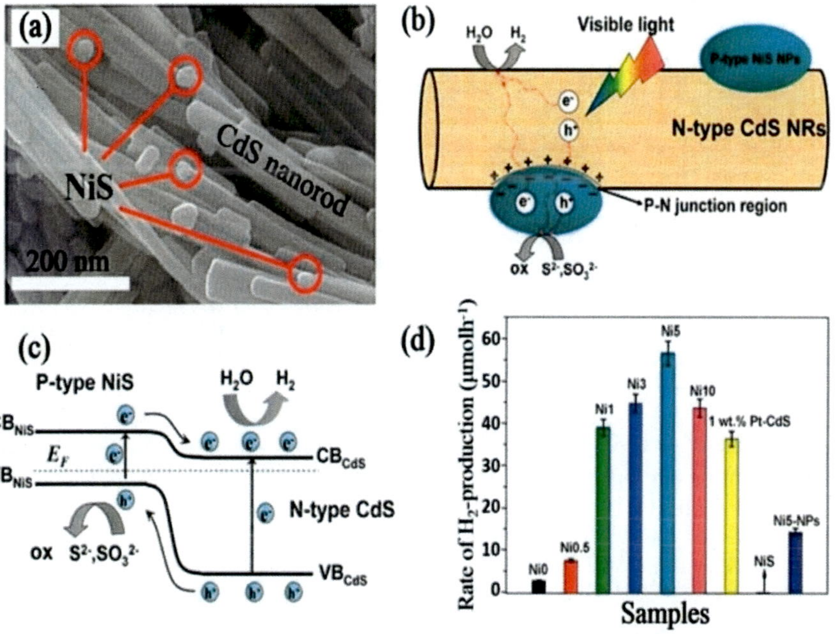

Fig. 4.5 **a** SEM morphology of CdS-NiS nanocomposite photocatalysts; **b**, **c** diagram of electron–hole pair representation with CdS/NiS nanocomposite photocatalyst; **d** comparison of photocatalysis performance of CdS with various NiS ratio [68]

4.1.2.2 ZnS

Zinc sulfide is a good semiconducting material used for splitting of water extensively because of nanoscale morphology, unique electrical, and optical properties [72, 73]. It displayed high photocatalytic performance without any assistance of co-catalysts. The major problem which restricts its practical applications is low activation when exposed to visible light in photocatalysis due to its wide band gap energy of 3.6 eV. Therefore, different strategies such as doping, dye sensitization, and composite development with other semiconductors were applied to make ZnS visible-light-driven photocatalytic material to split water [74–76]. It was observed that the metal or nonmetal ion doping could result into the development of doping states within band gap of material, which ultimately lower its band gap. For example, nonmetal ions doping such as C, B, N, F, S, F, and Cl showed excellent performance via creating impurity doping states nearer to valence band in band gap of ZnS as displayed in Fig. 4.6 [75]. Lee et al. [76] observed band gap energies between 2.58 and 2.74 eV for N-TiO_2 photocatalyst, clearly representing modified optical properties after nitrogen doping. Similarly, other strategies were also found helpful to increase the photocatalysis efficiency of ZnS like dye sensitization, where dye could capture visible light and transport electrons directly to the conduction band of semiconductor, leading to well separation of charge carriers [74]. Moreover, composite formation with other

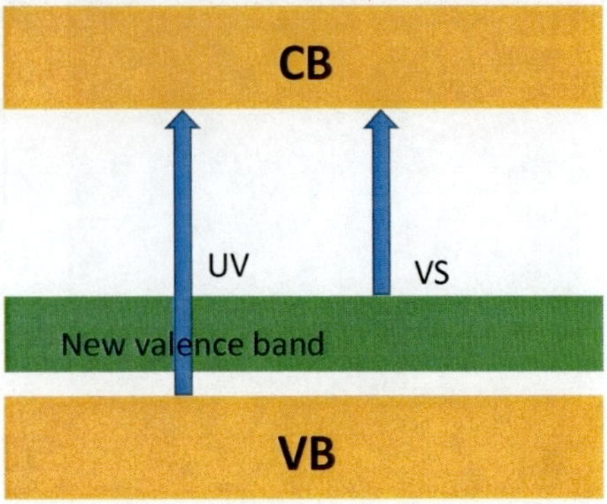

Fig. 4.6 Valence band formation within band gap of semiconducting material because of nonmetal ion doping

semiconductors is also efficient technique to overcome the recombination of charge carriers during photocatalysis. Therefore, the modification of ZnS by these methods is highly effective to utilize visible light for higher performance.

4.1.2.3 MoS_2

Molybdenum disulfide (MoS_2) was extensively applied in photocatalysis field for hydrogen generation by water splitting reaction due to exceptional electrical, optical, and photocatalytic characteristics. MoS_2 is a semiconductor material formed via the S-Mo-S bond, and layers are held together in stalks via van der Walls forces. The band gap of MoS_2 can vary from ~1.29 to ~1.8 eV which depends on number of layers [2, 77]. Earlier work in the 1970s on bulk MoS_2 proposed that this semiconductor would not be a particularly active catalyst for H_2 generation [78]. Nevertheless, since the development of MoS_2 nanostructures with good photocatalysis hydrogen evolution activity, research in this field is going on [79]. Two effective techniques were developed to achieve superior photocatalysis performance of MoS_2 in H_2 generation including increase in light absorption ability and higher charge transport properties [80, 81]. In order to increase light response range of MoS_2, it was modified by doping with metal/nonmetal ions and coupling with other narrow band gap semiconductors [82–85]. Previous reports have shown that a photocatalysis performance of MoS_2 could be improved by developing composite with other co-catalyst-semiconductors. The most used co-catalysts are $CdSe/MoS_2$, CdS-graphene-MoS_2, MoS_2-graphene, p-MoS_2/n-rGO, MoS_2/TiO_2, and MoS_2/g-C_3N_4 [86–92]. Liu et al. prepared MoS_2-CdS p–n junction-based heterostructure films having higher PEC performance for H_2 generation when visible photo-irradiated, and the IPCE of the photocatalyst was significantly enhanced due to development of p–n junction at interface of contact. Novel MoS_2/CdS photocatalyst not only extended the use of solar light but also inhibited the CdS photocorrosion [88].

It is greatly motivating that single metal sulfide photocatalyst performance is made excellent by simple modifications as proved in last decade. However, there are still few remaining chal-lenges in photocatalysis field to achieve novel photocatalysts to be used at commercial level.

4.1.2.4 Bi-Metal Sulfide

Bi-metal sulfide-based materials have got a lot of atten-tion as highly active photocatalysts for H_2 generation. For improving H_2 generation efficiency and avoiding the photocorrosion of metal sulfides that are the common problem of many photocatalysts which comprise sulfur, a S^{2-}/SO_3^{2-} solution [93]. Typically, the S^{2-} ions respond quickly when they react with photocreated positive charge carriers to generate polysulfides, decreasing overall photocatalytic efficiency [63]. The composite photocatalytic system consisting of more than two semiconductors is common technique to get higher efficiency of nanophotocatalyst [94]. Few commonly developed bi-metal sulfide materials include Zn (Cu)-In-S, Zn-Ga-S, Cu-Ga- S (Se), and Cd-Zn-S [95–99].

Among these, a novel solid solution of $Cd_{1-x}Zn_xS$ designed from CdS and ZnS has shown a great potential because of controlled band structure and outstanding photocatalysis efficiency in producing H_2 gas from splitting of water [3]. Many synthesis methods have been utilized to prepare $Cd_{1-x}Zn_xS$ solid solution including co-precipitation, microemulsion, and hydrothermal/solvothermal [100–103]. Xing et al. [104] prepared $Cd_{1-x}Zn_xS$ solid solution using facile co-precipitation technique following calcination under nitrogen environment. They found higher quantum performance of $Cd_{0.62}Zn_{0.16S}$ when illuminated with ultraviolet and visible light. The further modification in $Cd_{1-x}Zn_xS$ bi-metal sulfide especially by graphene was found effective for enhanced H_2 production. For example, Zhang et al. report highlights the efficient role of rGO in promoting H_2 production photocatalysis activity of $Zn_{0.8}Cd_{0.2}S$ nanoparticles [105]. As described in Fig. 4.7a, b, effective role of rGO was due to good electron accepting and transporting properties which separated the photocreated charge carriers based on photocatalyst ($Zn_{0.8}Cd_{0.2}S$) system. In this study, RGO-$Zn_{0.8}Cd_{0.2}S$ heterostructure photocatalyst displayed an improved hydrogen generation activity of approximately 4.5 times than that of Pt $-$ $Zn_{0.8}Cd_{0.2}S$ using similar reaction conditions (Fig. 4.7c), representing the excellence of rGO loaded $Zn_{0.8}Cd_{0.2}S$ among metal sulfide photocatalysts.

Besides these performances, it was noted that photocatalysis efficiency of bi-metal sulfide-based materials is still low which require further improvement. Moreover, the conventional hydrothermal synthesis of these nanomaterials is generally time taking and expensive. Therefore, facile synthesis of metal sulfide-based nanomaterials with improved photocatalysis performance still remains a challenge.

4.1.2.5 Multi-metal Sulfide

Multi-metal sulfide based on superior physical, chemical, and electronic characteristics is also promising for applications in photocatalysis of water splitting [106]. It is

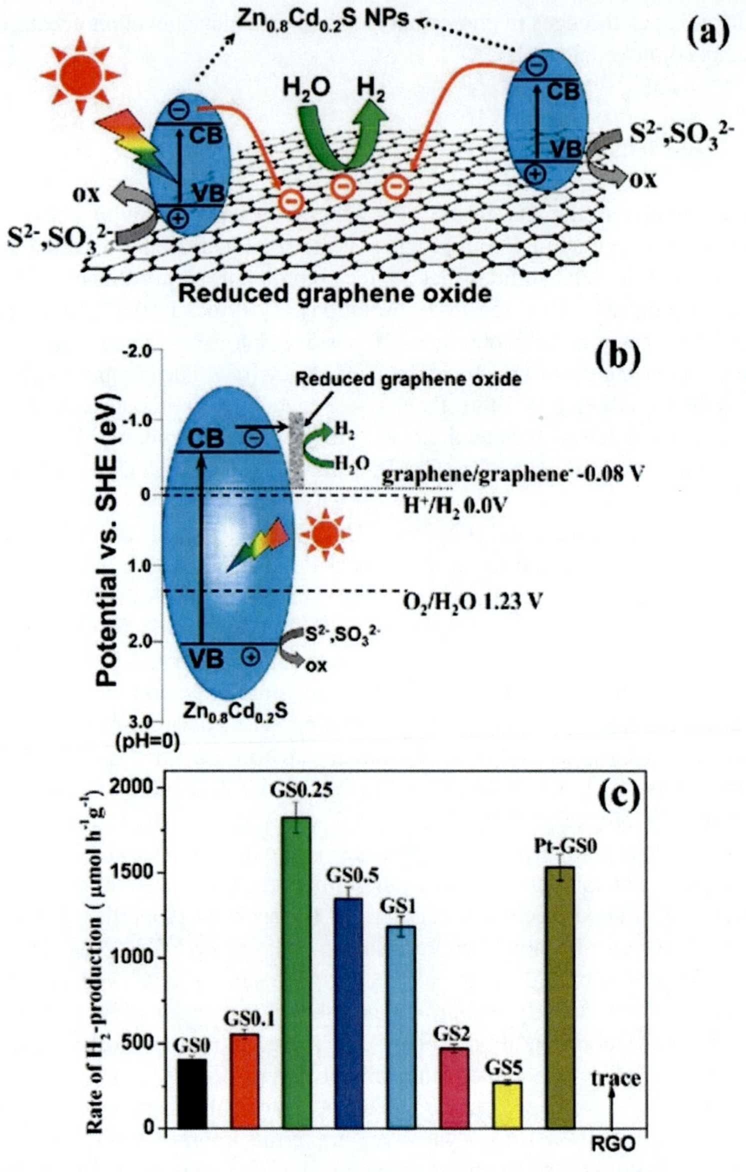

Fig. 4.7 **a** Schematically represented charges transportation in the RGO-$Zn_{0.8}Cd_{0.2}S$ nanocomposites under solar illumination; **b** proposed reaction mechanism for the improved electron transport in RGO-$Zn_{0.8}Cd_{0.2}S$ composite for H_2 generation under solar illumination; **c** H_2 generation rates of RGO-$Zn_{0.8}Cd_{0.2}S$ nanocomposites with various ratios of RGO. Pt-GS0: GS0 incorporated with the optimum 1 wt% of Pt co-catalyst under solar illumination [105]

a most common technique to develop a visible-light active composite photocatalyst via mixing different nanomaterial having similar crystal structure but varying band gap energy using chemical methods [107]. The band gaps of nanomaterials can be adjusted by varying solution proportions of different input products. But it was also observed that different types of structures might cause lower crystallinity of resulting composite materials, which may show lower performance of photo-catalyst. Moreover, by using semiconductors with narrow and wider band gaps, a variety of materials can be obtained, which could display superior efficiencies compared to individual nanomaterials [108]. Therefore, multi-metal sulfide-based photocatalysts showing higher light absorption and photocatalytic activity were designed by researchers.

$CuInS_2$ was also researched for splitting of water to produce hydrogen. However, $CuInS_2$ in photocatalytic water splitting displayed lower efficiency due to wider band gaps. Therefore, $CuInS_2$ was modified for improving its capability to produce hydrogen via photocatalytic decomposition of water [109, 110]. In a study, Yu et al. used gallium to increase $CuInS_2$ conduction band edge for efficient activation of water splitting reduction reaction [109]. Superior H_2 generation rate was achieved for $CuIn_{0.3}Ga_{0.7}S_2$ with reduction rate of about 750 m mol g^{-1} h^{-1}. Another semiconductor material based on Cu_2ZnSnS_4 has been studied which showed high yield of hydrogen during photocatalytic hydrogen production [111, 112].

Besides, above-mentioned compounds, some other compounds including Cu_2NiSnS_4, Cu_2CoSnS_4, and Au/Cu_2FeSnS_4 were also reported in photocatalytic water splitting for H_2 production [113, 114]. In conclusion, compared with other semiconductor photocat-alysts, the multi-metal sulfide displayed superior performances in photocatalysis H_2 generation and it can be expected that more these types of nanomaterials will be synthesized and applied for photocatalysis applications.

4.1.3 Metal-Free-Based Nanomaterials

The toxic metal-leaching due to photocorrosion was considered to be substantial problem for metal-based photocatalysts regarding practical applications. Therefore, the metal-free semiconductor such as graphitic carbon nitride ($g\text{-}C_3N_4$) has a huge attraction as a H_2 producing photocatalyst nanomaterial in splitting of water application [115, 116]. Graphitic carbon nitride can be prepared by process of thermal condensation using N_2-containing precursors such as cyanamide, dicyandiamide, melamine, thiourea, and urea (Fig. 4.8) [117]. In addition, $g\text{-}C_3N_4$ has a band gap energy approximately 2.7 eV thus could capture visible light effectively [118, 119]. Moreover, the suitability of its band structure for oxidation and reduction reactions make it an attractive nanomaterial to be used in photocatalysis and photochemical processes to produce energy [120, 121]. Besides these advantages, $g\text{-}C_3N_4$ has some limitations when it is applied in water splitting reactions via photocatalysis. Majorly, its wider band gap, low charges mobility, and less separation ability for electrons and holes during photocatalysis are the main drawbacks [122]. Different strategies were developed to enhance its performance in photocatalysis such as morphological

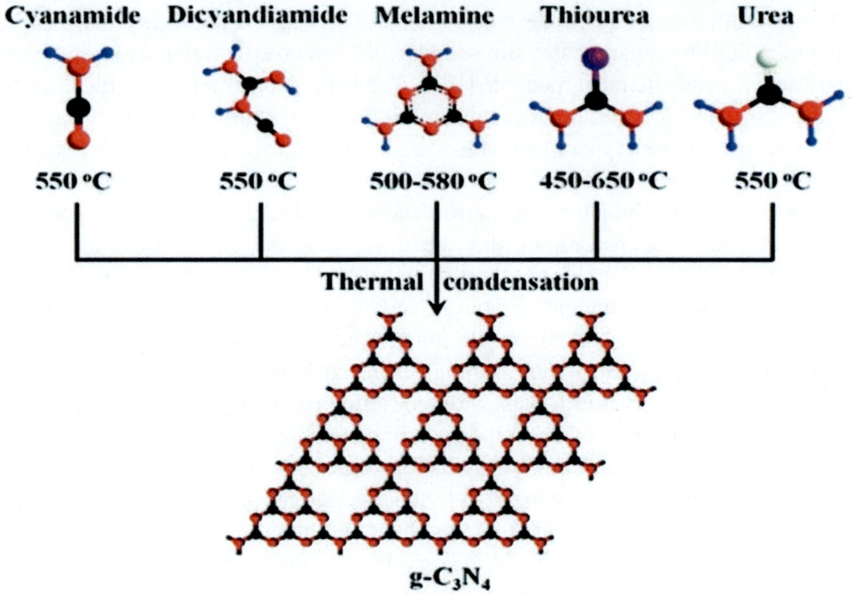

Fig. 4.8 Diagram for major preparation method of bulk g-C3N4 the color codes used in representation of chemical phase structures: C, black; N, red; H, O, white blue; S, purple [117]

modification, heterostructure formation, and incorporation of doping elements into g-C_3N_4 [123–125].

There are reports which demonstrate an increase in photocatalysis activity for H_2 generation when lower thickness g-C_3N_4 material. For example, thinner g-C_3N_4 nanosheets having thickness of 3 nm were previously fabricated with the use of urea precursor [124].

This sheet-like nanostructure has favored transport process charges and achieved higher H_2 generation. In another report, free-standing g-C_3N_4 nanosheets developed by Yang et al. which displayed good hydrogen production performance [126]. The H_2 generation rate over synthesized g-C_3N_4 nanosheets was observed to be 93 mmol h^{-1}, representing nine times higher activity compare to bulk g-C_3N_4 under visible photo-illumination.

Zhao et al. [127] developed hollow nanospheres of mesoporous g-C_3N_4 (MCNHN) by applying simple deposition technique and observed increased H_2 evolution when exposed to visible photo-illumination. Figure 4.9 displays good photocatalysis H_2 production by MCNHN under visible-light illumination. The achieved hydrogen evolution rate over MCNHN sample was 659.8 μ mol g^{-1} h^{-1}, which is 22-fold compare to pristine g-C_3N_4 (29.6 μ mol g^{-1} h^{-1}). This superior activity of this novel nanostructure was specially attributed to higher surface area which provided more reaction sites on the surface. Furthermore, this photocatalyst also helped in good transportation and separation of charges, thereby, higher rate of H_2 generation via photocatalysis was achieved. Figure 4.9b represents schematic

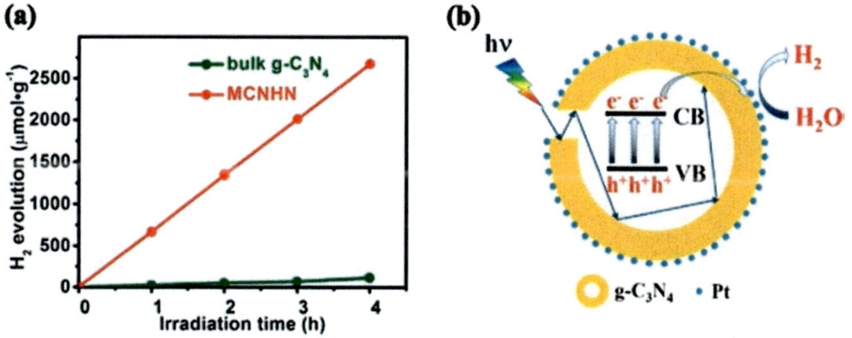

Fig. 4.9 **a** Time plot of hydrogen generation and **b** a mechanistic presentation for photocatalysis H_2 production on MCNHN [127]

mechanistic diagram of photocatalysis reaction for H_2 production over MCNHN. Upon light absorption by active sites of MCNHN, the photocreated negative electrons were transported to CB from VB and then transferred to platinum present on MCNHN surface. In this way, the recombination of charge carriers was greatly reduced, and as a result, photocatalytic hydrogen production was enhanced with this novel photocatalyst.

Bulk g-C_3N_4 even with low dimensionality has less ability to absorb visible light, higher charge carrier's recombination properties, thus, there are many studies to make it an excellent photocatalyst. The g-C_3N_4/semiconductor-based heterostructures had showed enhanced separating ability of charges, and therefore, superior photocatalytic activity [128–130]. Moreover, g-C_3N_4-based heterojunction can be developed using a smaller band gap semiconducting nanomaterial and a larger band gap via coupling process [131, 132]. Zhang et al. [133] investigated the preparation of g-C_3N_4-TiO_2 heterostructure material with higher H_2 generation rate when irradiated with visible light.

Another way to increase activity of g-C_3N_4 in photocatalysis to generate higher H_2 is to dope it with metal or nonmetal element [118, 134]. It is known that anion doping is an important scheme to modify basic structure of g-C_3N_4 and to enhance photocatalysis activity via favoring reaction on material surface. Feng et al. [135] studied the P and P-doped g-C_3N_4 nanostructures as visible-light active materials for hydrogen generation reaction. The P@P-g-C_3N_4 exhibited increased optical absorption, higher transfer, and separation of photocreated charges. When carbon atoms were substituted by P atoms (Fig. 4.10a) in crystal structure of g-C_3N_4, then extra negative charges were aligned into a π-conjugated triazine-ring and created a positive charge P^+, thus favoring the fast separation of the photocreated excited electrons.

Moreover, band gap variation between phosphorous (P) and P-doped g-C_3N_4 tends to an increase in photocatalytic activity (Fig. 4.10b). It was perceived that photo-excited negative charges of P-doped g-C_3N_4 were transferred to phosphorous via direct conduct because conduction band edge of g-C_3N_4 is approximately 1.2 V versus NHE and it is more negative compare to P (−0.25 V vs. NHE). This

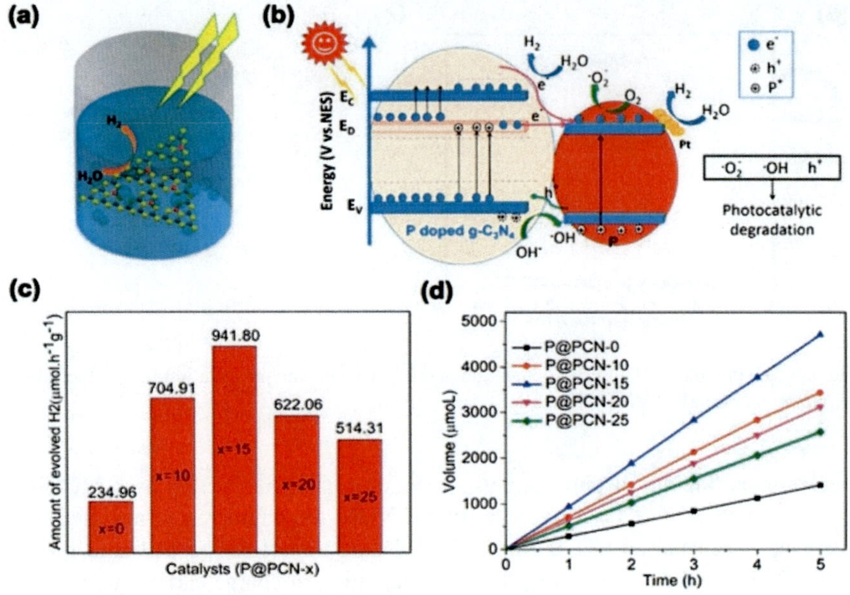

Fig. 4.10 **a**, **b** Schematically illustrated reaction mechanism indicating H_2 production by the P@P-g-C_3N_4 catalyst; **c** comparison of the generation rates of H_2; and **d** hydrogen production rate of the P@P-g-C_3N_4 nanocomposites [135]

difference of band potential creates a contact and induce electric field. Therefore, electrons present on the phosphorous surface can be easily taken by oxygen, resulting into lower recombination of charge carriers. Among all samples, the P@P-g-C_3N_4-15 showed the superior photocatalytic activity and hydrogen production rare was 941.80 μmol h^{-1} g^{-1}, which is about fourfold than that of conventional g-C_3N_4 as shown in Fig. 4.10c, d.

4.2 Nanomaterials for Photocatalytic CO_2 Conversion

The nanostructured materials have attracted attention recently because nanomaterials have an incredibly higher surface area, structural fine-tuning, and low cost. In these days, an enormous progress has been made about new nanostructures such as metal-based and metal oxide-based photocatalysts for CO_2 conversion into useful hydrocarbons or fuels. Therefore, in this section, the recent development on nanomaterials is briefly summarized including an overview of recent developments summarizing the most researched nanostructured materials including most researched TiO_2-based, Cu-based, and g-C_3N_4-based nanomaterials applied for photocatalytic CO_2 conversion application are highlighted.

4.2.1 TiO_2-Based Nanomaterials

Metal oxide semiconductor such as titanium dioxide was extensively utilized as photocatalytic material in CO_2 conversion due to some superior properties like nontoxicity, chemical, and biological stability and easy accessibility [136–138]. TiO_2 has a good conduction and valence band potentials to start photocatalysis for CO_2 reduction. However, its band gap is higher which restricts its use under solar light irradiation. Thus, a lot of attempts were tried to expand visible-light absorptivity as well activity of TiO_2 photocatalyst. Firstly, TiO_2 modification with doping elements such as nitrogen, sulfur, and carbon is a highly good strategy to shorten its band gap for higher absorption in visible-light range [139–142]. Li et al. reported a nanomaterial (N-doped mesoporous TiO_2) which showed increased photocatalysis efficiency for carbon dioxide conversion to methane (CH_4) than that of TiO_2 alone under the same conditions due to extended visible-light absorption [143]. Besides this, TiO_2 modified with transition metal ions is also an efficient technique to extend its visible-light absorption range [144, 145]. In a study, loading of 1.5 wt% Ni^{3+} in TiO_2 matrix has increased absorption in visible-light range as well as sufficiently improving CO_2 photo reduction [145].

Plasmonic metal nanoparticles including Au and Ag nanoparticles loaded on semiconductor photocatalyst are thought to display some positive influences such as enhancing the charge carrier's separation via the development of Schottky barriers, extending the visible-light absorption, and increasing surface electron generation by their surface plasmon resonance (SPR) influences [146, 147]. Many researchers have reported the plasmonic Au, Ag, and Cu nanoparticles incorporated TiO_2 nanomaterials for photocatalysis conversion of CO_2 with enhanced visible-light active photocatalysis efficiency. In a report, Cronin et al. [148] investigated TiO_2/Au nanomaterial with improved performance (24-fold higher than pure TiO_2) for CO_2 photoreduction. They attributed the enhanced photocatalytic activity to high electric fields induced by SPR of gold nanoparticles. Yu et al. utilized facile silver mirror technique to fabricate Ag/TiO_2 nanocomposite which exhibited an effective photocatalytic performance due to surface plasmonic resonance (SPR) effects [149]. During photocatalytic conversion of CO_2, different chemical compounds were observed such as CH_4 and CH_3OH along with few traces of C_2 and C_3 species including acetone and acetaldehyde. In another report, copper incorporated TiO_2 nanoflowers films (TNF) for improved photocatalysis reduction of carbon dioxide into methanol were studied by [150]. They observed that the prepared copper modified TNF with 0.5 m mol ratio of Cu^{2+} displayed higher methanol production under ultraviolet–visible and/or only ultraviolet photo-illumination, which was about 6 and 3.6 times superior than pristine TNF. Superior CH_3OH production was majorly ascribed to well charge separation due to the development of local surface plasmon resonance (LSPR). The LSPR was actually induced by Cu nanoparticles which provided hot electrons to the photocatalytic CO_2 reduction process for methanol yield as shown in possible reaction mechanism of Fig. 4.11.

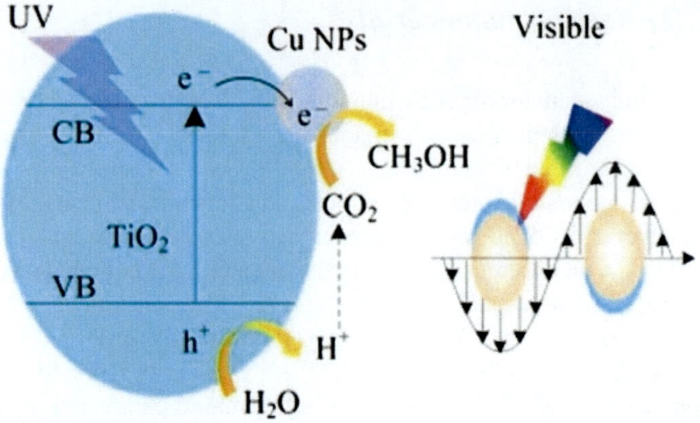

Fig. 4.11 Mechanistic illustration of the process involved in the photocatalysis reduction of CO_2 into CH_3OH by Cu-loaded TNF films [150])

One-dimensional (1D) TiO_2-based nanomaterials including nanotubes, nanorods, and nanowires were thoroughly investigated in field of photocatalysis. The main advantages of 1D TiO_2 nanomaterials consist of higher specific surface areas, good charge transfer, enhanced adsorption capability, and extended visible-light absorption because of light scattering and trapping effect [151]. A reasonable research has been conducted on the synthesis of 1D TiO_2-based photocatalytic materials with a major emphasis on CO_2 reduction to hydrocarbon and/or fuels. A variety of strategies includes heterostructure development with other visible-light responsive semiconductors and graphene coupling for achieving the increased photocatalytic CO_2 conversion rate. Li et al. studied a novel heterojunction comprising of Cu_2O nanoparticle loaded TiO_2 nanotube (TNT) arrays [152]. They used conventional electrochemical anodization method to prepare TNT arrays, while nanoparticles of Cu_2O were decorated using copper salt via electro-deposition technique. The Cu_2O/TNT fabricated with electro-deposition time of 0.5 h exhibited higher CH_4 production when exposed to visible photo-illumination. Although, the Cu_2O/TNT composite fabricated with 15 min of electro-deposition time displayed highest yield of CH_4 under solar simulated photo-irradiation. Improved efficiency of material was due to the following reasons: (i) TNT arrays provided good charge transfer rate and light absorptivity, (ii) optimal loading amount of Cu_2O led to enhanced visible-light activity, and (iii) the matched band edges of Cu_2O and TNT arrays well favored in separation of charges. Most possible reaction mechanism is shown in Fig. 4.12a. Xin et al. reported CdS and Bi_2S_3 modified TiO_2 nanotubes (CdS-TNT and Bi_2S_3-TNT) heterostructures [153]. The Bi_2S_3-TNT heterojunction photocatalyst exhibited a higher photocatalytic performance for CO_2 conversion giving 224.6 μmol/L yield of CH_3OH after 5 h light exposure than that of CdS-TNT (159.5 μmol/L) and pure TNT (102.59 μ mol/L) photocatalysts. It might be due the reason that the conduction band potentials of both CdS and Bi_2S_3 are higher than CB potential of TiO_2; thus, with photo-illumination, the photocreated electrons can be certainly transferred to

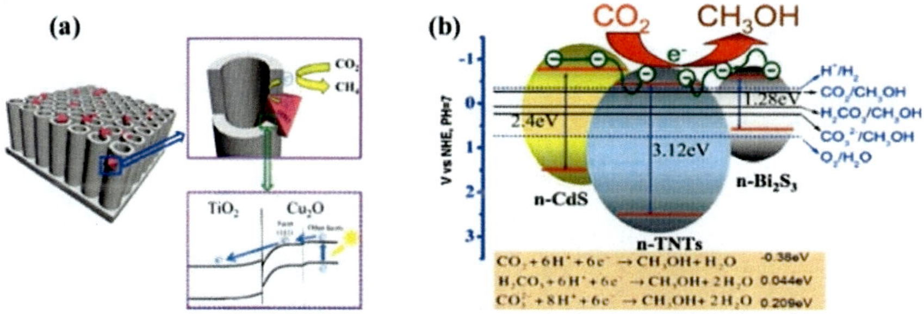

Fig. 4.12 Mechanisms illustrated for photocatalysis reduction process of CO_2 into: **a** CH_4 over Cu_2O nanoparticles decorated TNT (reproduced with permission from Ref. [152]). **b** CH_3OH employing CdS and Bi_2S_3 TNT photocatalytic material (reproduced with permission from Ref. [153])

CB of TiO_2, leading to lower reunion of charges and increased photocatalytic efficiency. The authors proposed a photocatalytic reaction mechanism for band gap configuration and carbon dioxide reduction to methanol as shown in Fig. 4.12b.

The last ten years' research confirmed that there is an exponential increase in the development of graphene-modified semiconducting materials composites for photocatalysis field [154]. A lot of research on graphene and its derivative-based nanostructures has been extensively done [155–157]. Especially the incorporation of graphene into TiO_2 gives benefits simultaneously, such as increased reactant absorptivity, better charge transportation and separation, and an improved photo-absorption range [158]. Recently, Zhao et al. [159] have reported Pt/TiO_2 core–shell photocatalyst material covered by ultrathin sheets of rGO prepared via a self-assembly process (Fig. 4.13a). This ternary nanocomposite showed a higher CH_4 formation rate of 41.3 μmol g^{-1} h^{-1} along with a good selectivity of 99.1%. Figure 4.13b demonstrates the CO_2 photocatalytic reduction mechanism with H_2O to CH_4 employing (Pt/TiO_2)-rGO composite. All-solid-state electron multiple transmitting scheme was very effective for the vectorial electron transport in TiO_2/Pt/rGO, which showed good charge separating performances. In another study, platinum incorporated graphene loaded reduced blue titanium dioxide (Fig. 4.13c) has provided much higher photocatalysis production for C_2H_6 (77 μ mol g^{-1}) and CH_4 (259 μmol g^{-1}) under photo-illumination of seven hours [160]. Figure 4.13d displays the charge transfer mechanism occurring in this outstanding ternary composite nanomaterial.

Rambabu et al. [161] prepared rGO/TiO_2 multi-leg nanotubes (MLNTs) via wrapping GO/rGO sheets on TiO_2 nanotubes and utilized for photocatalysis reduction of carbon dioxide. Such a novel nanostructure has capability to separate photogenerated charge carriers and increase their transportation during photocatalysis of CO_2 conversion. The rGO/TiO_2-based nanocomposite exhibited higher photocatalytic efficiency with highest CO production rate of 1348 μ mol g^{-1} in just 20 min (Fig. 4.14a). Yeh et al. [162] also observed similar results when they used GO as co-catalyst along with TiO_2 to reduce carbon dioxide. The formation rate of carbon monoxide (CO) in graphene supported TiO_2 nanocrystals was much higher than pristine TiO_2.

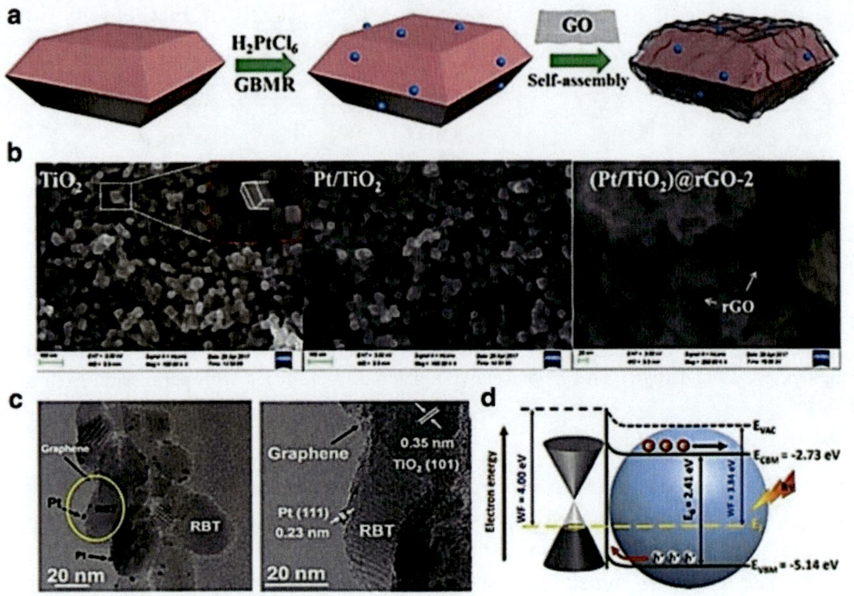

Fig. 4.13 **a** Schematics of two-step preparation of (Pt/TiO_2)-rGO-based photocatalyst materials. **b** SEM images of TiO_2, Pt/TiO_2, and (Pt/TiO_2)-rGO samples, Reprinted from [159] with permission from Elsevier, copyright (2018), **c** HRTEM images of Pt-loaded graphene-wrapped reduced blue titanium dioxide, **d** schematically illustrated diagram of energy level positions at the graphene-RBT contact. Reproduced from [160] with permission from RSC, copyright (2018)

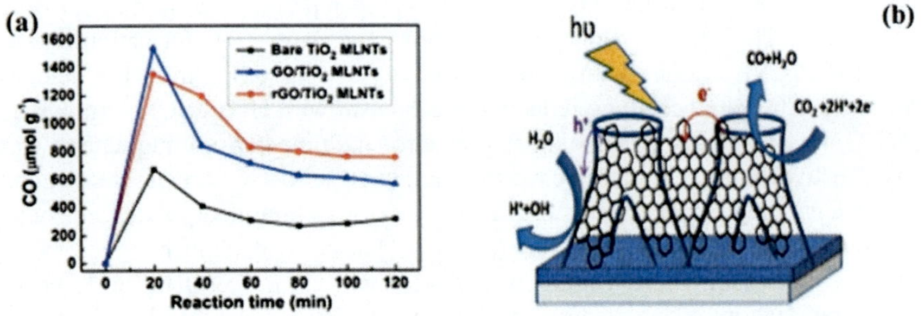

Fig. 4.14 **a** Time depending product of CO output as a result of photocatalysis conversion of carbon dioxide employing on the surface of photocatalytic material, **b** illustration displays the photocatalysis CO reduction procedure involved in GO/rGO incorporated TiO_2 multi-leg nanotubes. Reprinted from [161] with permission from Elsevier, copyright (2019)

Moreover, the interconnected sheets of GO-rGO between neighboring nanotubes could capture and transport photo-excited electrons which ultimately increase CO_2 conversion rate. This is illustrated in schematic diagram of Fig. 4.14b.

4.2.2 Copper-Based Nanomaterials

Copper-based semiconducting materials such as copper oxide (CuO), cuprous oxide (Cu_2O), and cuprous sulfide (Cu_2S) with smaller band gap energies of approximately 1.7 eV, 2.2 eV, and 1.2 eV, respectively, are particularly important because of their excellent visible-light absorbing ability, appropriate conduction band (CB) potential, and good choosiness toward carbon dioxide conversion into useful fuels (Fig. 4.15) [163]. However, sometimes, their smaller band gap results into lower photocatalysis performance because of speedy reunion of photocreated charges. Therefore, the modification of copper-based nanomaterials with n-type semiconducting materials to design p–n heterostructure is a useful approach for enhanced photocatalytic CO_2 conversion into hydrocarbons. This strategy can successfully increase the lifetime of the photogenerated electrons and holes by inducing electric field at interface contact of photocatalyst between both semiconducting materials, which is highly beneficial to increase the efficiency of CO_2 photocatalytic reduction [164].

For an example, α-Fe2O$_3$/Cu_2O nanocomposites were reported for the CO_2 photocatalytic conversion application [165]. A higher CO yield of about 1.67 μ mol g^{-1} h^{-1} under visible photo-illumination was achieved, which was significantly superior than that of pure α-Fe_2O_3 and/or Cu_2O. They credited higher photocatalysis activity to the formation of p–n heterostructures.

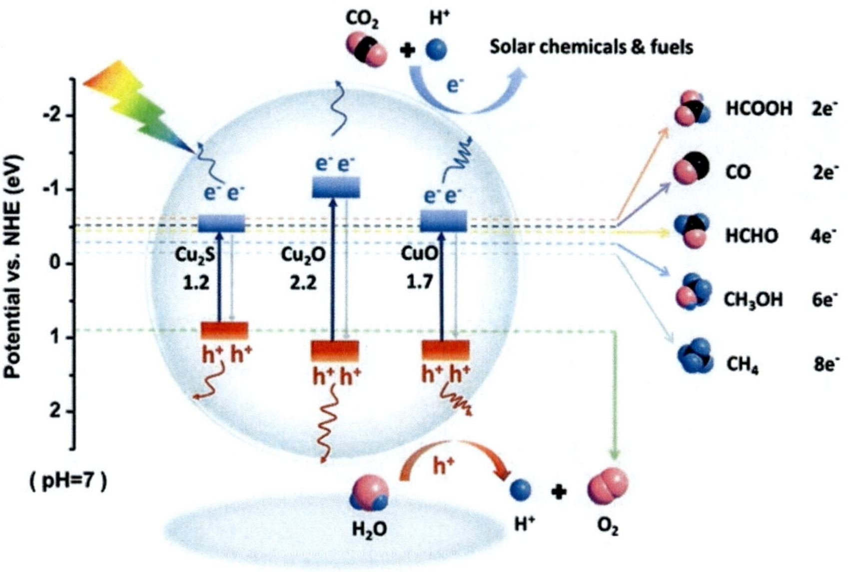

Fig. 4.15 Schematically representation of photoreduction of CO_2 to useful fuels and chemicals over Cu-based materials. Reprinted from Ref. [163], Copyright (2019), with permission from RSC

Another efficient method to reduce CO_2 via photocatalysis comprises coupling of different semiconductor materials by an all-solid-state Z-scheme heterostructure method [166]. Z-scheme-based heterojunction system provides good interfacial charge transfer between two semiconductor photocatalysts. For example, $BiVO_4$ nanoparticles were fabricated on C-coated copper oxide nanowires arrays (BVO/C/Cu_2O NWAs) using sequential ionic-layer adsorption reaction (SILAR) method to design Z-scheme system which provided efficient charge flow through the composite interface and showed higher performance for photocatalysis CO_2 reduction in H_2O presence under solar photo-illumination (Fig. 4.16a) [167]. The carbon interlayer between BVO and Cu_2O was highly useful because it protected copper oxide from photocorrosion and also acted as an electron mediator due to good electron mobility in heterojunction system. Therefore, overall it leads to more effective separation of photocreated charges and consequently increased photocatalytic activity for CO_2 conversion. As displayed in Fig. 4.16b, the copper is shielded via very dense BVO/C/Cu_2O NWAs thoroughly with great homogeneity. However, Fig. 4.16c demonstrates that heterostructure is consists of separate BVO/C/Cu_2O nanowires having very rough surface. Furthermore, Fig. 4.16d shows the performance of all photocatalysts where ternary BVO/C/Cu_2O NWAs exhibited higher photocatalysis

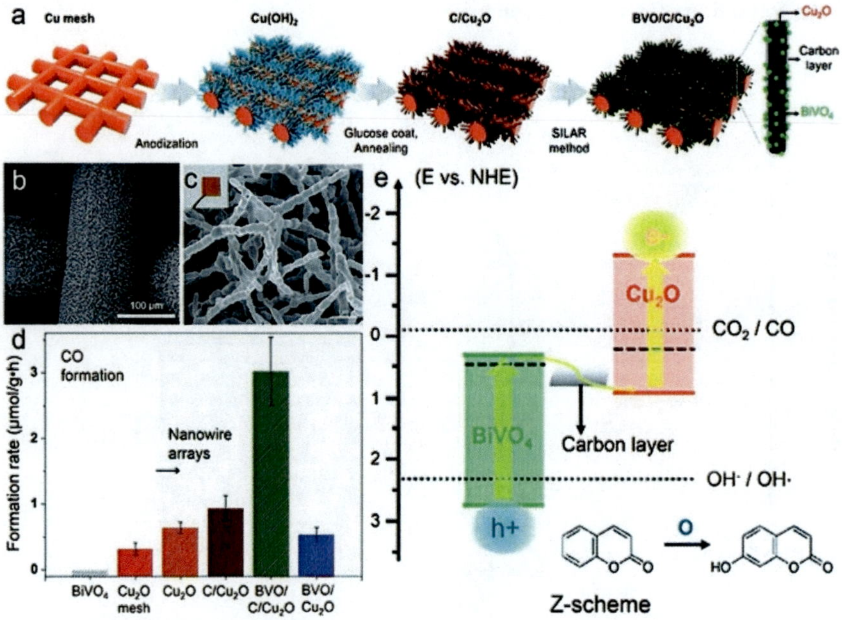

Fig. 4.16 **a** Synthesis process scheme of BVO/C/Cu_2O NWAs; **b** SEM image of BVO/C/Cu_2O nanowires arrays; **c** SEM image of BVO/C/Cu_2O nanowires arrays with inset displaying their optical image; **d** CO generation rate for all composite; **e** representation of Z-scheme charge transfer in BVO/C/Cu_2O nanowires arrays for photocatalysis of CO_2 conversion. Reproduced from Ref. [167] with permission from ACS, copyright (2019)

activity for CO_2 conversion into carbon monoxide due to the growth of Z-scheme heterostructure. Figure 4.16e demonstrates that the photogenerated electrons were efficiently transferred from BVO conduction band to C and then transported to copper oxide valence band. In this way, photocreated holes in copper oxide were utilized effectively besides resulting in self-photocorrosion.

Importantly, C containing materials (carbon nanotubes, activated carbon, and graphene) are valuable for effective electron–hole separation and transport due to their higher conductivity and superior electron mobility [140]. They also provide higher specific surface area which makes their suitability to form composites with one-dimensional Cu-based nanomaterials for achieving superior efficiency in CO_2 reduction reactions. For this purpose, Khatri et al. [168] have designed rGO-CuO heterojunction by decorating CuO nanorods on rGO nanosheets using chemical method. This novel heterojunction based on rGO-CuO demonstrated increased performance as photocatalyst under visible-light illumination and also showed good stability in aqueous medium. Among all samples, rGO-CuO heterostructure showed higher production of CH_3OH (1228 μ mol g^{-1}), which is sevenfold compare to pristine CuO nanorods as displayed in Fig. 4.17a. Moreover, the reusability of the optimal rGO-CuO-116 photocatalysts is represented in Fig. 4.17b. The results show that photocatalytic performance for the production of CH_3OH remains approximately similar even after six successive cycles of reaction which clearly confirm that this material has good stability. Authors attributed the excellent performance of rGO-CuO-based photocatalyst to good interaction between rGO sheets and copper oxide which supported the effective separation and transfer of charge carriers as shown in Fig. 4.17c.

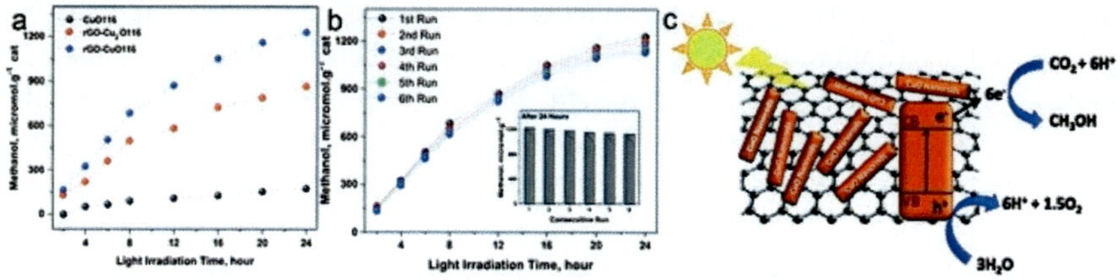

Fig. 4.17 **a** Photocatalytic process of CO_2 reduction vs. photo-irradiation time employing different photocatalytic materials to yield CH_3OH; **b** reusability of rGO-CuO116 for the CH_3OH production via photocatalysis conversion of CO_2 as a function of light illumination time (inset represents CH_3OH production after 24 h of photocatalysis for six consecutive runs); **c** schematically represented plausible reaction mechanism of photocatalysis reduction of CO_2 into methanol over rGO/CuO nanocomposites under solar photo-illumination. Reprinted from Ref. [168], Copyright (2016), with permission from Elsevier

4.2.3 g-C_3N_4-Based Nanomaterials

Graphitic carbon nitride (g-C_3N_4) is visible-light responsive photocatalytic material which has various benefits like good sustainability, simple preparation, and cost-effectiveness when compared with other semiconducting nanomaterials in carbon dioxide reduction processes [169–171]. In particular, g-C_3N_4 has an intrinsic band gap of approximately 2.7 eV, therefore, it gives good visible-light harvesting capability in higher wavelength region of solar spectrum [118]. The g-C_3N_4 supports superior CO_2 reduction performance due to its appropriate conduction band potential typically situated at ~1.3 V (vs. NHE at pH = 7), which is highly suitable for driving multi-electron transport processes to generate fuels like CO, HCOOH, and CH_3OH [140]. Firstly, g-C_3N_4 was applied in photocatalysis of CO_2 conversion in 2011 by Zhang et al. [172]. After that, it attracted a huge number of researchers to use g-C_3N_4 as an ideal and economical photocatalyst because it can be easily synthesized by simple chemical techniques [117]. On the other hand, it has several shortcoming including lower electrical conductivity, low separation rate of photocreated charges, and no visible photo-absorption in higher wavelength region as reported by many researchers [173, 174].

Elemental doping is an outstanding technique to design highly visible-light responsive g-C_3N_4 like photocatalytic materials with higher performance because it is simple in fabricating and advancing crystal structure of g-C_3N_4. Previously, researchers confirmed that incorporation of different metal ions impurities including Li, K, Mg, Cu, and Fe into g-C_3N_4 has greatly improved photocreated charges separation and advanced optoelectronic characteristics of this material [175–180]. For example, Shi et al. [177] recently studied an enhanced photocatalytic activity with higher yield of about 49.45 μmol g^{-1} for CO production using Cu-decorated g-C_3N_4 under visible-light illumination of 5 h. The authors attributed the superior photocatalytic performance to extended visible-light absorption, promoted electron mobility, and inhibited charge carrier's recombination due to Cu incorporation into g-C_3N_4. In another report of Tang et al. [179], Mg was doped into the g-C_3N_4 by an in situ hydrothermal process. As the main atom of chlorophyll, magnesium doping showed good effects on CO_2 adsorption capability and photogenerated charge separation efficiency of g-C_3N_4 by acting as an active site. The Mg/g-C_3N_4 showed a superior visible-light active CH_4 production yield of 17.09 μmol g^{-1} along with CO yield of 4.13 μmol g^{-1} during photocatalytic CO_2 reduction.

Furthermore, Sun et al. [180] synthesized K-modified g-C_3N_4 using post-treatment of urea polymerized g-C_3N_4 by KOH as the chemical activation source. In this way, the resulting nanomaterial experienced a depletion of carbon in g-C_3N_4 motifs, leading to the development of carbon vacancies along with an amine-rich functionalized surface. The amino groups were the sources of increase adsorption of CO_2 adsorption onto g-C_3N_4, whereas the K and carbon vacancies acted in synergy to favor charge transfer in g-C_3N_4. This novel photocatalyst (K/g-C_3N_4) showed 5.4 and 5.6-time higher performance in CO and CH_4 productions, respectively, as compare to than that of the undoped g-C_3N_4.

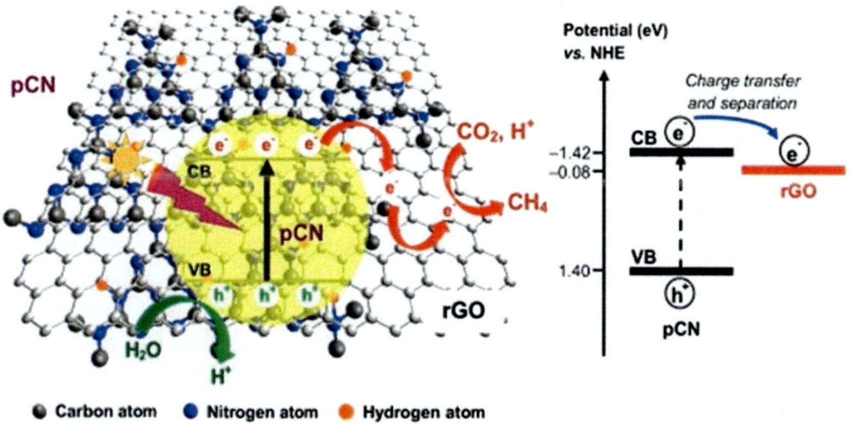

Fig. 4.18 Schematically illustrated charge transportation and separation over g-C_3N_4/rGO composite for carbon dioxide conversion using water and methane mix solution under visible light. Reprinted from Ref. [182], Copyright (2015), with permission from Elsevier

Furthermore, g-C_3N_4 modified with carbon nanomaterials including carbon dots and reduced graphene oxide (rGO) has also attained greater attention [181–183]. For example, Ong et al. [183] observed an improved photocatalytic CO_2 reduction efficiency of g-C_3N_4/carbon dots. The total CH_4 and CO generation rates by the g-C_3N_4/3 wt% carbon dot composite were found to be 29.23 and 58.82 μmol g^{-1} visible-light irradiation (10 h), which were 3.6 and 2.28 times superior, respectively, than that of g-C_3N_4 alone. In another study, Chai et al. [182] prepared a 2D/2D hybrid heterojunction nanocomposite using rGO and protonated g-C_3N_4. The g-C_3N_4/rGO nanocomposite photocatalysts with various rGO ratios exhibited increased photocatalytic activity for CO_2 reduction to CH_4 as compared to pure g-C_3N_4. The sample (g-C_3N_4/15 wt% rGO) provided a CH_4 production amount of 13.93 μmol g^{-1} over 10 h with a quantum efficiency of 0.56% and higher stability. They claimed that the surface modification of g-C_3N_4 increased the 2D/2D interlamination region and increased electrons-holes separating ability of material (Fig. 4.18), therefore, increasing the photocatalysis conversion efficiency.

4.3 Nanomaterials for Photocatalytic N_2 Fixation

A variety of nanomaterials have been synthesized by different techniques to enhance their performance for N_2 reduction to ammonia. Due to strong dependency of physical/chemical properties of nanostructure materials on their shape, size, and specific surface area, the rational design is of great importance to further promote sustainable N_2 fixation activity. In the following section, we will highlight the current research progress on novel and efficient nanomaterials used for photocatalytic reduction of N_2 reduction to ammonia.

4.3.1 TiO_2-Based Nanomaterials

The excellent properties of TiO_2 proved it the best choice for photocatalytic applications [184–186]. First time in 1988, Bourgeois et al. [187] have reported the pure TiO_2 for photocatalytic N_2 fixation application after heating in air atmosphere because thermal process can generate surface defects which lower band gap of TiO_2 by creating impurity states within the band gap. For instance, photocatalysis conversion of nitrogen to NH_3 was achieved over TiO_2 by generating oxygen vacancies on surface of material by Shiraishi et al. [188]. Among all samples, JRC-TiO-6 (rutile phase) displayed excellent N_2 fixation performance, exhibiting a 2.7-time higher efficiency when 2-PrOH was utilized as a sacrificial electron donor in photocatalytic reaction. They observed that active sites for N_2 reduction are the Ti^{3+} species on the oxygen vacancies, which acted as adsorption sites for N_2 and trapping sites for the photogenerated conduction band electrons. These excellent properties promoted the efficient reduction of N_2 to NH_3. The solar into chemical energy conversion performance was found to be 0.02%, which was higher than early reported photocatalysts.

The photocatalytic system containing oxygen vacancies (OVs) is helpful for N_2 fixation activity because excessive OVs in TiO_2 can increase the adsorption capacity of N_2 molecules as well as improved charge carrier's separation efficiency. Zhang et al. [189] recently optimized OVs concentration in TiO_2 which enhanced the charge separation performance by three times and showed higher activation toward N_2 molecules. Figure 4.19a, b displays the N_2 fixation (Fig. 4.19a) and normalized N_2 fixation rate (Fig. 4.19b) of anatase and reduced TiO_2. They found normalized N_2 photofixation rate to about 324.86 μmol h^{-1} g^{-1} (under full spectrum) and the corresponding apparent quantum yield (AQY) under 365 nm to ~1.1%, which were higher than the other reported literature. Moreover, they observed that NH_4^+ was the only product without any other product like N_2H_4 and H_2. In addition to this, the performance remains nearly constant after six cycles of photocatalytic experiment, showing that R-340 has a good photostability under full spectrum irradiation (Fig. 4.19c). Figure 4.19d demonstrates the proposed schematic diagram of photocatalytic reaction for N_2 conversion, where it can be seen that oxygen vacancies played their role to enhance the efficiency of TiO_2.

Schrauzer et al. [190] synthesized Fe-doped TiO_2 using varying Fe ratios from 0% to 1% by heat treatment of iron(III) sulfate-impregnated anatase TiO_2. They observed that this newly developed photocatalyst demonstrated a great ability to reduce N_2 under UV illumination. Interestingly, ammonia along with a small quantity of hydrazine was detected in their work and ammonia yield reached a maximum value with 0.2% Fe_2O_3-doped TiO_2. Augugliaro et al. utilized a similar impregnation-calcination technique to prepare Fe_2O_3/TiO_2 composite supported by γ-alumina

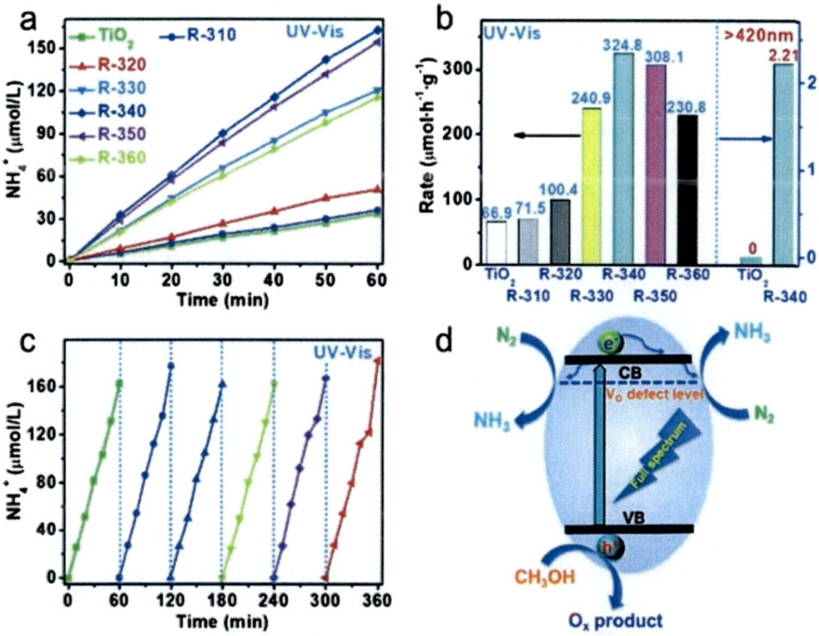

Fig. 4.19 Full spectrum N_2 conversion (**a**) and normalized N_2 fixation rate (**b**) of anatase and reduced TiO_2. The recycle ability evaluation for N_2 fixation for R340 under full spectrum illumination (**c**). The schematically drawn photocatalysis N_2 conversion (**d**). The 50 mg of samples were mixed into 100 mL of 10 vol% methanol solution under full spectrum and visible photo-irradiation for N_2 fixation, respectively Reproduced from Ref. [189] with permission from RSC, copyright (2018)

[191]. Superior NH_3 generation rates were obtained in gas-solid fluidized bed reactor in which the reactants were nitrogen and water. In conclusion, the authors attributed that the doping TiO_2 with Fe ions favored the well separation of photocreated electron–hole pairs because no ammonia production was observed with pure TiO_2 under UV illumination.

Moreover, the plasmonic metal modification of TiO_2 is another efficient technique for improving photo-absorption and photocatalysis activity for NH_3 preparation [192–194]. Au nanoparticles display a good surface plasmonic absorption at 550 nm of visible region. Wang et al. reported that Au nanoparticles incorporated TiO_2 nanosheets (ultrathin) exhibited effective N_2 photocatalytic reduction to NH_3 under visible-light illumination [195]. In conclusion, a greater number of research efforts were made to the bulk or surface modification of TiO_2 photocatalysts in order to improve the photocatalytic performance for ammonia preparation. However, TiO_2-based photocatalysts still suffer from low photocatalytic efficiencies for the synthesis of ammonia; therefore, more efforts into synthetic techniques that can overcome these challenges are required in this field.

4.3.2 Bi-Based Nanomaterials

Bismuth-based semiconductors are good choice as photocatalysts due to good optical properties because of narrow band gap and good photocatalytic efficiency for nitrogen reduction to ammonia [185, 196]. Typically, bismuth oxyhalides are ternary compound semiconductors and are generally defined by the general formula $[Bi_lOmXn]$, where $X = F$, Cl, Br, and I [197]. They have a layered structure which can provide suitable space for the polarization of atoms and form internal electric field for well separation of charge carriers [186, 198]. It is known that the oxygen vacancies (OVs) are common surface anion defects on oxide surfaces having low formation energy. Li et al. prepared ternary oxide material of BiOBr, where OVs effected photocatalytic N_2 reduction in two ways [196]. Firstly, the presence of OVs reduced the charge carrier's recombination, thus favoring photoexciton interaction toward fast charge carrier's production via trapping hot electrons as confirmed by higher emission of defect sights in PL spectroscopy (Fig. 4.20a, b). In the nitrogen environment, PL emission peak was greatly reduced because of interfacial transport of electrons from BiOBr to N_2, which was due to important role played by oxygen vacancies for enhanced N_2 activation. Oxygen vacancies supported the generation of un-saturated centers for direct nitrogen adsorption, and prominent electron-rich centers with localized electrons able to activate N-N triple bond via a charge back

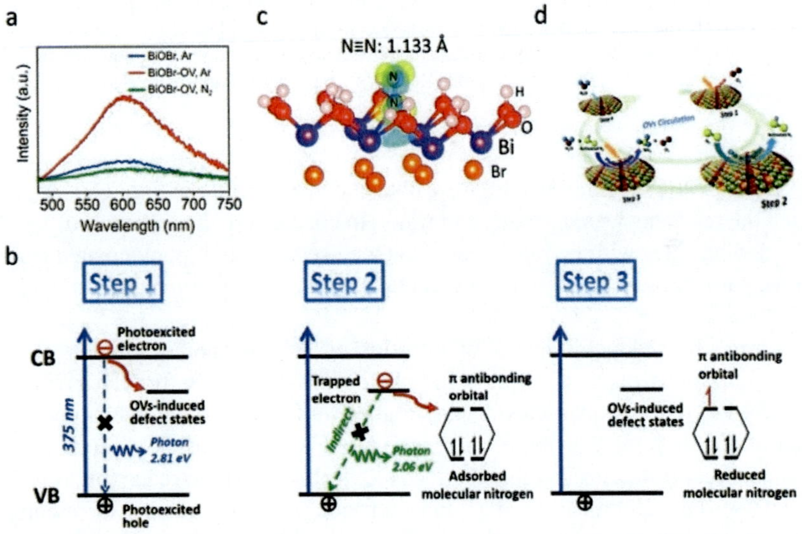

Fig. 4.20 **a** Steady photoluminescence spectra of the as-prepared BiOBr in diverse environment. **b** Schematically illustrated diagram for OVs-assisted interfacial charge transport for N_2 reduction. **c** Terminal end-on adsorption structure of N_2 on oxygen vacancy of BiOBr. **d** Schematically drawn diagram for representing photochangeable OVs production for N_2 fixation over Bi5O7Br. Reproduced with permission from ref. [196], with permission from ACS, copyright (2015) and reproduced from Ref. [199] with permission from Wiley, copyright (2017)

donation (Fig. 4.20c). The density functional theory (DFT) calculation also confirmed N_2 activation via enhanced N-N bond-length of 1.133 Å between the N-N triple bond (1.078 Å) and double bond (1.201 Å). A similar N_2 activation scheme could also be realized on the OVs of Bi_5O_7Br by Ye et al. [199], which is a good derivative of BiOBr having visible-light responsive changeable OVs production characteristic (Fig. 4.20d).

In addition to the bismuth oxyhalides, other bismuth-based nanomaterials including BiO, Bi_2MoO_6, and $Bi_2O_2CO_3$ were also used for photocatalytic synthesis of ammonia [200–202]. BiO quantum dots with an average size in the range of 2–5 nm showed superior photocatalytic activity for nitrogen reduction reaction under simulated solar light [200]. It was observed that the low-valence surface Bi(II) species were highly beneficial as active sites for N_2 activation. Moreover, Bi_2MoO_6 were also efficient photocatalysts for ammonia synthesis from ordinary air [201]. The authors attributed the N_2 activation capability of Bi_2MoO_6 nanocrystals to edge-exposed coordinately un-saturated Mo atoms in Mo-O polyhedra. Additionally, 2D bismuth oxyhalides and related bismuth-based photocatalysts have served as model nanomaterial for exploring the influences of OVs, facet-engineering, and band gap modification for photocatalytic N_2 conversion to NH_3 [185]. Particularly, combining experimental data and theoretical simulations is an effective approach for achieving good photocatalytic properties in this field.

4.3.3 *Metal Sulfide-Based Nanomaterials*

It is known that transition metal sulfides carry smaller band gaps and also appropriate bands potentials for nitrogen reduction reaction (NRR). Moreover, their superior optical characteristics make them effective choice for the photocatalysis nitrogen conversion reactions to synthesize NH_3 [186, 203]. In 1980, Miyama et al. fabricated CdS/Pt binary wafer nanocomposites and applied them for heterogeneous photocatalysis to reduce N_2 into high yield ammonia than that of pure CdS [204]. Furthermore, ammonia synthesis via photocatalysis was obtained under visible photo-illumination over CdS/Pt/RuO_2 ternary composite [205]. It was found that the N_2 molecules got activation via reacting with the photo-excited Ru complexes and RuO_2 was responsible for trapping photogenerated electrons from the conduction band of CdS. Although metal sulfides display excellent visible-light absorption properties; however, the photocorrosion of photocatalyst is a serious concern during photocatalytic ammonia synthesis. For obtaining higher photocatalysis activity for N_2 fixation to produce ammonia, Ni_2P was loaded on metal sulfide which captured more photocreated charges and also increase their transportation [206].

A lot of organic-sulfide-based materials were investigated for N_2 fixation because of interdisciplinary research in different fields. Brown et al. studied nitrogen reduction to prepare NH_3 with high generation rate by the MoFe protein modified by CdS nanorods to prepare biohybrid complexes [207]. In this biohybrid, CdS nanorods were utilized for photosensitization of MoFe protein in order to achieve photon

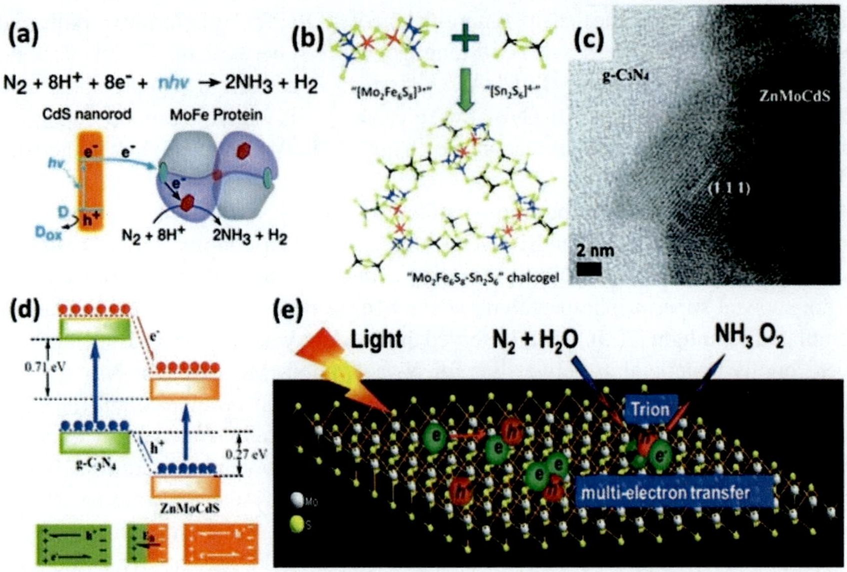

Fig. 4.21 **a** Reaction mechanism for nitrogen conversion to ammonia employing CdS: MoFe protein bio-hybrids. Reprinted with permission [207] Copyright 2016, AAA of Science, **b** schematically illustrated $Mo_2Fe_6S_8$–Sn_2S_6 biomimetic chalcogel composition. Reprinted with permission. [208] Copyright 2015, ACS, **c** HRTEM display of g-C_3N_4/ZnMoCdS. **d** Charge transport phenomenon at the heterostructure contact of g-C_3N_4/ZnMoCdS. Reproduced with permission. [210] Copyright 2016, RSC. **e** Schematically showed multi-electron N_2 conversion reaction with trion. Reprinted with permission. [211] Copyright 2017, Elsevier

energy from adenosine 5′-triphosphate (ATP) as shown in Fig. 4.21a. Moreover, the yield of ammonia was similar as generation by physiological via nitrogenase with energy supplied by ATP because active sites of nitrogenase consist of elements like Fe, Mo, and S. In another report, Banergee et al. reported the biomimetic chalcogels containing FeMoS inorganic clusters which could easily reduce N_2 to NH_3 in aqueous solution under light illumination [208]. The $Fe_2Mo_6S_8$ chalcogel was almost similar to the MoFe active site in the enzyme and was attached by Sn_2S_6 ligands to design an amorphous network (Fig. 4.21b). They found that that the structural similarity of nitrogenase could be well-functional and have superior characteristics compare to nitrogenase. Moreover, some researchers observed that the sulfur vacancies were beneficial for enhanced photocatalytic activity. For example, Hu et al. designed $Zn_{0.1}Sn_{0.1}Cd_{0.8}S$ having S-vacancies and utilized for fixation of nitrogen when exposed to photo-illumination [209]. Interestingly, S-vacancies presented more adsorption sites on material surface that led to higher fixation of nitrogen molecules. Moreover, the sulfur vacancies also captured photo-excited electrons, therefore, improving the separation of charges. A hybrid of g-C_3N_4/ZnMoCdS (Fig. 4.21c) was also prepared with effective transportation properties of charges. The photo-excited negative charges had been transported from g-C_3N_4 to the ZnMoCdS material, while positive charges were moved in reverse way (Fig. 4.21d) [210]. In this way, an enhanced visible-light absorption after hybridization was achieved for better photocatalysis efficiency. The photocatalytic conversion of nitrogen to ammonia with ultrathin layer MoS_2 prepared under different preparation conditions was studied by Sun et al. [211]. They observed that few-layer molybdenum disulfide material produced greater quantity of ammonia. This was ascribed to the development of excitons in MoS_2 layers which trapped more electrons [212]; therefore, it was considered that charged excitons in MoS_2 acted as electron-rich species and hence higher performance in for N_2 fixation (Fig. 4.21e) [211].

4.3.4 Carbonaceous Nanomaterials

Carbonaceous photocatalysts have gained a lot of attention in field of photocatalysis since the development of boron-doped diamond in 2013 [213, 214]. Nowadays, the research focus is on carbonaceous nanomaterials for photocatalytic N_2 reduction activity. As described previously, g-C_3N_4 having N_2 vacancies has developed an emerging field in materials science since 2015 [116]. It was found that nitrogen vacancies (NVs) are suitable in chemisorption and fixation process of nitrogen. Furthermore, NVs have the capability to increase separation ability of charges via capturing photo-excited negative charges and supporting electron transport to the adsorbed N_2 molecules [215]. Furthermore, the self-modification of g-C_3N_4 without fabricating hybrids was also an effective approach for altering electronic structure of g-C_3N_4, thus enhancing photocatalytic nitrogen reduction [216]. Xue et al. developed porous few-layer g-C_3N_4 (PFL-g-C_3N_4) using a facile molecular self-assembly process [217]. The ultrathin PFL-g-C_3N_4 photocatalyst displayed higher nitrogen fixation efficiency of 8.20 m M h^{-1} $gCat^{-1}$ as compare to pristine g-C_3N_4 (2.92 m M h^{-1} g Cat^{-1}). The photocatalytic N_2 reduction performance was observed higher under all conditions of sacrificial reagents as shown in Fig. 4.22a. The N_2 conversion schematic representation over PFL-g-C_3N_4 photocatalyst is displayed in Fig. 4.22b. The reunion of photoproduced charges is largely inhibited in this novel photocatalyst because of porous nature of material. Additionally, a few layers' ultrathin nanostructure provided greater surface area of material and thereby extra surface active sites for higher N_2 reduction.

In addition to above, modifications of g-C_3N_4 with iron and ruthenium incorporation have been extensively reported [219, 220]. For example, H-termination of ruthenium (Ru) incorporated g-C_3N_4 demonstrated good participation during fixation of N_2 molecules [221]. The heterostructure formation of g-C_3N_4 with other components is also an important approach for improved photocatalysis activity for N_2 reduction into ammonia. For instance, a highly efficient heterostructure based on g-C_3N_4 nanosheets/Bi_2MoO_6 (NCN/BMO) with varying percentage of bismuth molybdate was prepared by Elham et al. [218]. The NCN/BMO (30%) sample exhibited superior ability to prepare ammonia up to 3271 μmol L^{-1} g^{-1} that was 9.21-fold higher as compare to BMO alone, under mild conditions. The authors calculated

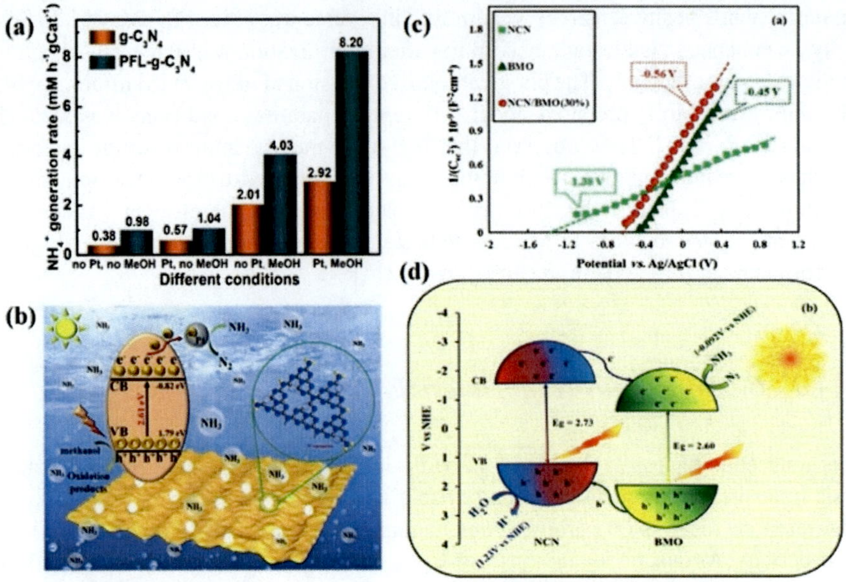

Fig. 4.22 **a** Photocatalytic N_2 fixation using g-C_3N_4 and PFL-g-C_3N_4 under varying conditions, **b** photocatalysis nitrogen fixation mechanism of PFL-g-C_3N_4, Reproduced from Ref. [217] with permission from Elsevier, copyright (2020), **c** the Mott–Schottky graphs for NCN, BMO, and NCN/BMO (30%) photocatalysts, **d** proposed reaction mechanism for the nitrogen conversion capability of the NCN/BMO composites when exposed to solar light. Reproduced from Ref. [218] with permission from Elsevier, copyright (2020)

the flat-band potential (E_{fb}) for all prepared semiconducting materials as shown in Fig. 4.22c. It can be seen that slope these plots is positive confirming that these nanomaterials are n-type semiconductors, where the position of conduction band is ~0.10 V more negative compare to E_{fb} [222]. Based on this, a possible mechanistic reaction for improved N_2 fixation over NCN/BMO photocatalyst is presented in Fig. 4.22d. The band gap NCN/BMO was decreased in synthesized composites via aligning nanostructures of both semiconductors, leading to higher visible-light absorption. Furthermore, this novel binary photocatalyst also suppressed the recombination of charge carriers due to good conductivity of Pt present at the surface of catalyst.

References

1. J. Zhu, D. Chakarov, M. Zäch, Nanostructured materials for photolytic hydrogen production, in *Energy Efficiency and Renewable Energy Through Nanotechnology, Green Energy and Technology*, ed. by L. Zang. https://doi.org/10.1007/978-0-85729-638-2_13
2. X. Zong, C. Li, Photocatalytic water splitting on metal oxide-based semiconductor photocatalysts. 09ijhgfds

3. J. Shi, Y. Hara, C. Sun, M.A. Anderson, X. Wang, Three-dimensional high-density hierarchical nanowire architecture for high-performance pho-toelectrochemical electrodes. Nano Lett. **11**, 3413–3419 (2011)
4. Z. Pan, Y. Qiu, J. Yang, M. Liu, L. Zhou, Y. Xu, L. Sheng, X. Zhao, Y. Zhang, Synthesis of three-dimensional hyperbranched TiO_2 nanow-ire arrays with significantly enhanced photoelectrochemical hydrogen production. J. Mater. Chem. A **3**, 4004–4009 (2015)
5. Y. Li, H. Yu, W. Song, G. Li, B. Yi, Z. Shao, A novel photoelectro-chemical cell with self-organized TiO_2 nanotubes as photoanodes for hydro-gen generation. Int. J. Hydrogen Energy **36**, 14374–14380 (2011)
6. M. Altomare, M. Pozzi, M. Allieta, L.G. Bettini, E. Selli, H_2 and O_2 photocatalytic production on TiO_2 nanotube arrays: Effect of the anodization time on structural features and photoactivity. Appl. Catal. B Environ. **136–137**, 81–88 (2013)
7. M. Gao, L. Zhu, W.L. Ong, J. Wang, G.W. Ho, Structural design of TiO_2-based photocatalyst for H_2 production and degradation applications. Catal. Sci. Technol. **5**, 4703–4726 (2015)
8. Y. Pan, T. Zhou, J. Han, J. Hong, Y. Wang, W. Zhang, R. Xu, CdS quantum dots and tungsten carbide supported on anatase–rutile compos-ite TiO_2 for highly efficient visible-light-driven photocatalytic H_2 evolution from water. Catal. Sci. Technol. **6**, 2206–2213 (2016)
9. Y. Lai, J. Gong, C. Lin, Self-organized TiO_2 nanotube arrays with uniform platinum nanoparticles for highly efficient water splitting. Int. J. Hydrogen Energy **37**, 6438–6446 (2012)
10. Y. Ji, W. Guo, H. Chen, L. Zhang, S. Chen, M. Hua, Y. Long, Z. Chen, Surface Ti_{3+}/ Ti_{4+} redox shuttle enhancing photocatalytic H_2 production in ultrathin TiO_2 nanosheets/CdSe quantum dots. J. Phys. Chem. C **119**, 27053–27059 (2015)
11. X. Yang, W. Xuelian, J. Li, Y. Liu, TiO_2–Au composite nanofibers for photocatalytic hydrogen evolution. RSC Adv. **9**, 29097–29104 (2019)
12. D. Chen, K. Wang, D. Xiang, R. Zong, W. Yao, Y. Zhu, Significantly enhancement of photocatalytic performances via core–shell structure of ZnO@mpg-C_3N_4. Appl. Catal. B: Environ. **147**, 554–561 (2014)
13. Y. He, Y. Wang, L. Zhang, B. Teng, M. Fan, High-efficiency conversion of CO_2 to fuel over ZnO/g-C_3N_4 photocatalyst. Appl. Catal. B: Environ. **169**, 1–8 (2015)
14. S.T. Kochuveedu, Y.H. Jang, D.H. Kim, A study on the mechanism for the interaction of light with noble metal-metal oxide semiconductor nano-structures for various photophysical applications. Chem. Soc. Rev. **42**, 8467–8493 (2013)
15. A. Bhirud, S. Sathaye, R. Waichal, C. Park, B. Kale, In situ preparation of N-ZnO/graphene nanocomposites: Excellent candidate as a photocatalyst for enhanced solar hydrogen generation and high performance supercapaci-tor electrode. J. Mater. Chem. A **3**, 17050–17063 (2015)
16. J. Bai, Y. Li, R. Wang, K. Huang, Q. Zeng, J. Li, B. Zhou, A novel 3D ZnO/Cu_2O nanowire photocathode material with highly efficient photoelec-trocatalytic performance. J. Mater. Chem. A **3**, 22996–23002 (2015)
17. R. Krol, Principles of photoelectrochemical cells, in *Photoelectrochemical Hydrogen Production* (Springer, Berlin, 2012)
18. M. Li et al., Hierarchical nanowire arrays based on carbon nanotubes and Co3O4 decorated ZnO for enhanced photoelectrochemical water oxidation. J. Mater. Chem. A **3**(26), 13731–13737 (2015)
19. M. Sharma, D. Das, A. Baruah, A. Jain, A.K. Ganguli, Langmuir **30**, 3199–3208 (2014)
20. M. Higashi, K. Domen, R. Abe, J. Am. Chem. Soc. **134**, 6968–6971 (2012)
21. M.-Y. Tsang, N.E. Pridmore, L.J. Gillie, Y.-H. Chou, R. Brydson, R.E. Douthwaite, Adv. Mater. **24**, 3406–3409 (2012)
22. Y. Takahara, J.N. Kondo, T. Takata, D.L. Lu, K. Domen, Mesoporous tantalum oxide. 1. Characterization and photocatalytic activity for the overall water decomposition. Chem. Mater. **13**(4), 1194–1199 (2001)
23. L. Guo, H. Hagiwara, S. Ida, T. Daio, T. Ishihara, One-pot soft-templating method to synthesize crystalline mesoporous tantalum oxide and its photocatalytic activity for overall water splitting. ACS Appl. Mater. Interfaces. **5**, 11080–11086 (2013)

24. H. Cui, G. Zhu, Y. Xie, W. Zhao, C. Yang, T. Lin, G. Hui, F. Huang, Black nanostructured Nb_2O_5 with improved solar absorption and enhanced photoelectrochemical water splitting. J. Mater. Chem. A **3**, 11830–11837 (2015)
25. S. Ng, R. Wang, P. Zhang, B. Kang, D. Zhang, Z. Zhang, S. Qiu, Synthesis and visible-light photocatalytic performance of C-doped Nb_2O_5 with high surface area. Chem. Res. Chin. Univ. **34**(2), 274–278 (2018)
26. T. Yanagida, Y. Sakata, H. Imamura, Photocatalytic decomposition of H_2O into H_2 and O_2 over Ga_2O_3 loaded with NiO. Chem. Lett. **33**(6), 726–727 (2004)
27. Y. Sakata, Y. Matsuda, T. Yanagida, K. Hirata, H. Imamura, K. Teramura, Effect of metal ion addition in a Ni supported Ga_2O_3 photocatalyst on the photocatalytic overall splitting of H_2O. Catal. Lett. **125** (1–2), 22–26 (2008)
28. V.R. Eddy, D.W. Hwang, J.S. Lee, Photocatalytic water splitting over ZrO_2 prepared by precipitation method. Korean J. Chem. Eng. **20**, 1026–1029 (2003). https://doi.org/10.1007/BF02706932
29. Y.-S. Lai, H.-H. Lu, Y.-H. Su, High-efficiency water-splitting solar cells with low diffusion resistance corresponding to halochromic pigments interfacing with ZrO_2. ACS Sustain. Chem. Eng. **5**(9), 7716–7722 (2017)
30. M.B. Tahir, G. Nabi, N.R. Khalid, M. Rafique, Role ofeuropium on WO_3 performance under visible-light for photocatalytic activity. Ceramic Int. **44**, 5705–5709 (2018)
31. C.C. Mardare, A.W. Hassel, Review on the versatility of tungsten oxide coatings. Phys. Status Solidi A **216**, 1900047 (2019)
32. M.B. Tahir, N.R. Khalid, A.M. Asiri, Boosting the performance of visible light-driven WO_3/g-C_3N_4 anchored with $BiVO_4$ nanoparticles for photocatalytic hydrogen evolution. Int. J. Energy Res. **43**, 5747–5758 (2019)
33. L. Zhang, S. Li, B. Liu, D. Wang, T. Xie, Highly efficient CdS/WO_3 photocatalysts: Z-scheme photocatalytic mechanism for their enhanced photocatalytic H_2 evolution under visible light. ACS Catal. **4**, 3724–3729 (2014)
34. H. Katsumata, Y. Tachi, T. Suzuki, S. Kaneco, Z-scheme photocatalytic hydrogen production over WO_3/g-C_3N_4 composite photocatalysts. RSC Adv. **4**, 21405 (2014)
35. X. Liu, F. Wang, Q. Wang, Nanostructure-based WO_3 photoanodes for photoelectrochemical water splitting. Phys. Chem. Chem. Phys. **14**, 7894–7911 (2012)
36. O. Arutanti, A.B.D. Nandiyanto, T. Ogi, T. Kim, K. Okuyama, Influences of porous structurization and Pt addition on the improvement of photocatalytic performance of WO_3 particles. ACS Appl. Mater. Interfaces. **7**, 3009–3017 (2015)
37. J. Wang, N. Zhang, J.Z. Su, L.J. Guo, Alpha-Fe_2O_3 quantum dots: low-cost synthesis and photocatalytic oxygen evolution capabilities, RSC Adv. **6**(47), 41060–41066 (2016)
38. Q.J. Xiang, G. Chen, T.C. Lau, Effects of morphology and exposed facets of alpha-Fe_2O_3 nanocrystals on photocatalytic water oxidation, RSC Adv. **5**(64), 52210–52216 (2015)
39. J.A. Glasscock, P.R.F. Barnes, I.C. Plumb, A. Bendavid, P.J. Martin, Structural, optical and electrical properties of undoped polycrystalline hematite thin films produced using filtered arc deposition. Thin Solid Films **516**(8), 1716–1724 (2008)
40. Y. Ling, G. Wang, D.A. Wheeler, J.Z. Zhang, Y. Li, Sn-doped hematite nanostructures for photoelectrochemical water splitting. Nano Lett. **11**(5), 2119–2125 (2011)
41. G. Wang, Y. Ling, D.A. Wheeler, K.E.N. George, K. Horsley, C. Heske et al., Facile synthesis of highly photoactive alpha-Fe_2O_3-based films for water oxidation. Nano Lett. **11**(8), 3503–3509 (2011)
42. Y. Lin, S. Zhou, S.W. Sheehan, D. Wang, Nanonet-based hematite heteronanostructures for efficient solar water splitting. J. Am. Chem. Soc. **133**(8), 2398–2401 (2011)
43. R. Franking, L. Li, M.A. Lukowski, F. Meng, Y. Tan, R.J. Hamers et al., Facile postgrowth doping of nanostructured hematite photoanodes for enhanced photoelectrochemical water oxidation. Energy Environ. Sci. **6**(2), 500–512 (2013)
44. D. Xiong, W. Li, X. Wang, L. Liu, Nanotechnology **27**, 375401 (2016)
45. D. Zheng, X. He, W. Xu, X. Lu, Mater. Res. Bull. **96**, 354–359 (2017)

46. J. Qiu, H. Hajibabaei, M. R. Nellist, F. A. L. Laskowski, T. W. Hamann, S. W. Boettcher. ACS Cent. Sci. **3**, 1015–1025 (2017)
47. X. Yang, R. Liu, C. Du, P. Dai, Z. Zheng, D. Wang, A.C.S. Appl, Mater. Interfaces **6**, 12005–12011 (2014)
48. M. Hara, T. Kondo, M. Komoda, S. Ikeda, K. Shinohara, A. Tanaka et al., Cu_2O as a photocatalyst for overall water splitting under visible light irradiation. Chem. Commun. **34**(3), 357–358 (1998)
49. F. Wang, C.L. Li, F. Cao, F. Feng, Y.T. Zhao, W.F. Shangguan, A study of morphology-dependent Cu_2O nanocrystals for hydrogen evolution under visible light. J. Nanosci. Nanotechnol. **16**(12), 12555–12562 (2016)
50. G. Wang, R. van den Berg, C.D. Donega, K.P. de Jong, P.E. de Jongh, Silicasupported Cu_2O nanoparticles with tunable size for sustainable hydrogen generation. Appl. Catal. B **192**, 199–207 (2016)
51. I.V. Bagal, N.R. Chodankar, M.A. Hassan, A. Waseem, M.A. Johar, D-H. Kim, S-W. Ryu, Cu_2O as an emerging photocathode for solar water splitting—a status review. Inter. J. Hydrogen Energy **44**, 21351–21378 (2019)
52. R. Liu, Z. Zheng, J. Spurgeon, X. Yang, Enhanced photoelectrochemical water-splitting performance of semiconductors by surface passivation layers. Energy Environ. Sci. **7**, 2504–2517 (2014)
53. S. Ho-Kimura, S.J.A. Moniz, J. Tang, I.P. Parkin, A method for synthesis of renewable Cu_2O junction composite electrodes and their photoelectrochemical properties. ACS Sustain. Chem. Eng. **3**, 710–717 (2015)
54. Y. Yang, D. Xu, Q. Wu, P. Diao, Cu_2O/CuO bilayered composite as a high-efficiency photocathode for photoelectrochemical hydrogen evolution reaction. Sci. Rep. **6**, 35158 (2016)
55. T.W. Kim, K.-S. Choi, Nanoporous $BiVO_4$ photoanodes with dual-layer oxygen evolution catalysts for solar water splitting. Science **343**, 990–994 (2014)
56. K. Ye, H. Li, D. Huang et al., Enhancing photoelectrochemical water splitting by combining work function tuning and heterojunction engineering. Nat. Commun. **10**, 3687 (2019)
57. G. Dong, H. Hu, L. Wang, Y. Zhang, Y. Bi, Remarkable enhancement on photoelectrochemical water splitting derived from well-crystallized Bi_2WO_6 and $Co(OH)_x$ with tunable oxidation state. J. Catal. **366**, 258–265 (2018)
58. D.P. Sahoo, S. Patnaik, D. Rath, K.M. Parida, Synergistic effects of plasmon induced $Ag@Ag_3VO_4/ZnCr$ LDH ternary heterostructures towards visible light responsive O_2 evolution and phenol oxidation reactions, Inorg. Chem. Front. **5**, 879–896 (2018)
59. T. Kako, N. Kikugawa, J. Ye, Photocatalytic activities of $AgSbO_3$ under visible light irradiation. Catal. Today **131** (1–4), 197–202 (2008)
60. H. Kadowaki, N. Saito, H. Nishiyama, H. Kobayashi, Y. Shimodaira, Y. Inoue, Overall splitting of water by RuO_2-loaded $PbWO_4$ photocatalyst with d10s2-d0 configuration. J. Phys. Chem. C **111**(1), 439–444 (2006)
61. H. Kato, N. Matsudo, A. Kudo, Photophysical and photocatalytic properties of molybdates and tungstates with a scheelite structure. Chem. Lett. **33**(9), 1216–1217 (2004)
62. Y. Hosogi, Y. Shimodaira, H. Kato, H. Kobayashi, A. Kudo, Role of Sn21 in the band structure of SnM_2O_6 and $Sn_2M_2O_7$ (M_5Nb and Ta) and their photocatalytic properties. Chem. Mater. **20**(4), 1299–1307 (2008)
63. A. Kudo, Y. Miseki, Heterogeneous photocatalyst materials for water splitting. Chem. Soc. Rev. **38**, 253–278 (2009)
64. K. Zhangab, L. Guo, Metal sulphide semiconductors for photocatalytic hydrogen production, Catal. Sci. Technol. (2013). https://doi.org/10.1039/c3cy00018d
65. X. Ninga, G. Lu, Photocorrosion inhibition of CdS-based catalysts for photocatalytic overall water splitting. Nanoscale (2020). 10.1039/C9NR09183A
66. N.Z. Bao, L. Shen, T. Takata, K. Domen, Chem. Mater. **20**, 110 (2008)
67. M. Sathish, R.P. Viswanath, Catal. Today **29**, 421 (2007)
68. J. Low, J. Yu, M. Jaroniec, S. Wageh, A.A. Al-Ghamdi, Heterojunction photocatalysts. Adv. Mater. **29**, 1601694 (2017)

69. J. Zhang, S.Z. Qiao, L. Qi, J. Yu, Fabrication of NiS modified CdS nanorod p-n junction photocatalysts with enhanced visible-light photocatalytic H_2-production activity. Phys. Chem. Chem. Phys. PCCP **15**, 12088–12094 (2013)
70. Y.-J. Heo, Y. Zhang, K.Y. Rhee, S.-J. Park, Synthesis of PAN/PVDF nanofiber composites-based carbon adsorbents for CO_2 capture. Compos. Part B Eng. **156**, 95–99 (2019)
71. Y. Zhang, M. Park, H.Y. Kim, S.J. Park, Moderated surface defects of Ni particles encapsulated with NiO nanofibers as supercapacitor with high capacitance and energy density. J. Colloid Interface Sci. **500**, 155–163 (2017)
72. D. Chen, F. Huang, G. Ren, D. Li, M. Zheng, Y. Wang, Z. Lin, ZnS nano-architectures:photocatalysis, deactivation and regeneration. Nanoscale **2**, 2062–2064 (2010)
73. A.A. Ashkarran, Absence of photocatalytic activity in the presence of the photoluminescence property of Mn–ZnS nanoparticles prepared by a facile wet chemical method at room temperature. Mater. Sci. Semicond. Process. **17**, 1–6 (2014)
74. Gang-Juan Lee, Jerry J. Wu, Recent developments in ZnS photocatalysts from synthesis to photocatalytic applications—a review. Powder Technol. **318**, 8–22 (2017)
75. X. Chen, S. Shen, L. Guo, S.S. Mao, Semiconductor-based photocatalytic hydrogen generation. Chem. Rev. **110**, 6503–6570 (2010)
76. G.J. Lee, C.L. Hong, V. Batalova, G. Mokrousov, J.J. Wu, Synthesis of nitrogen-doped ZnS with camellia brushfield yellow nanostructures for enhanced photocatalytic activity under visible light irradiation. Int. J. Photoenergy **2013**, 1–7 (2013)
77. N.R. Khalid, M. Bilal, M.B. Tahir, M. Shakil, T. Iqbal, M. Rafique, N. Yousaf, S.S.A. Gillani, Interfacial coupling effect of Ag_2O nanorods over MoS_2 microflowers for improved photocatalytic activity. Ceramic Int. **46**, 6856–6859 (2020)
78. H. Tributsch, J.C. Bennett, Electrochemistry and photochemistry of Mos_2 layer crystals. 1. J. Electroanal. Chem. **81**, 97–111 (1977)
79. B. Hinnemann, P.G. Moses, J. Bonde, K.P. Jorgensen, J.H. Nielsen, S. Horch, I. Chorkendorff, J.K. Norskov, Biomimetic hydrogen evolution: MoS_2 nanoparticles as catalyst for hydrogen evolution. J. Am. Chem. Soc. **127**, 5308–5309 (2005)
80. R. van de Krol, Y. Liang, J. Schoonman, J. Mater. Chem. **18**, 2311–2320 (2008)
81. T. Thurston, J. Wilcoxon, J. Phys. Chem. B **103**, 11–17 (1999)
82. S. Wi, M. Chen, H. Nam, A.C. Liu, E. Meyhofer, X. Liang, Appl. Phys. Lett. **104**, 232103 (2014)
83. N.R. Khalid, Z. Israr, M.B. Tahir, T. Iqbal, Highly efficient Bi_2O_3/MoS_2 p-n heterojunction photocatalyst for H_2 evolution from water splitting. Int. J. Hydrogen Energy (2020)
84. X. Huang, Z. Zeng, S. Bao, M. Wang, X. Qi, Z. Fan, H. Zhang, Nat. Commun. **4**, 1444 (2013)
85. D. Merki, X. Hu, Energy Environ. Sci. **4**, 3878–3888 (2011)
86. J.L. Gunjakar, I.Y. Kim, J.M. Lee, Y.K. Jo, S.J. Hwang, Exploration of nanostructured functional materials based on hybridization of inorganic 2D nanosheets. J. Phys. Chem. C **118**, 3847–3863 (2014)
87. D.O. Sigle, L. Zhang, S. Ithurria, B. Dubertret, J.J. Baumberg, Ultrathin CdSe in plasmonic nanogaps for enhanced photocatalytic water splitting. J. Phys. Chem. Lett. **6**, 1099–1103 (2015)
88. Y. Liu, Y.X. Yu, W.D. Zhang, J. Phys. Chem. C **117**, 12949–12957 (2013)
89. S. Min, G. Lu, Sites for high efficient photocatalytic hydrogen evolution on a limited-layered MoS_2 cocatalyst confined on graphene sheets—The role of graphene. J. Phys. Chem. C **116**, 25415–25424 (2012)
90. F. Meng, J. Li, S.K. Cushing, M. Zhi, N. Wu, Solar hydrogen generation by nanoscale p–n junction of p-type molybdenum disulfide/n-type nitrogen-doped reduced graphene oxide. J. Am. Chem. Soc. **135**, 10286–10289 (2013)
91. Q. Liu, Z. Pu, A.M. Asiri, A.H. Qusti, A.O. Al-Youbi, Y. Sun, One-step solvothermal synthesis of MoS_2/TiO_2 nanocomposites with enhanced photocatalytic H_2 production. J. Nanopart. Res. **15**, 1–7 (2013)
92. Y. Tian, L. Ge, K. Wang, Y. Chai, Synthesis of novel MoS_2/g-C_3N_4 heterojunction photocatalysts with enhanced hydrogen evolution activity. Mater. Charact. **87**, 70–73 (2014)

93. Z. Zou, J. Ye, K. Sayama, H. Arakawa, Direct splitting of water under visible light irradiation with an oxide semiconductor photocatalyst. Nature **414**, 625–627 (2001)
94. E.A. Kozlova, S.V. Cherepanova, D.V. Markovskaya, A.A. Saraev, E.Y. Gerasimov, V.N. Parmona, Novel photocatalysts Pt/Cd1 − xZnxS/ ZnO/Zn(OH)2: Activation during hydrogen evolution from aqueous solu-tions of ethanol under visible light. Appl. Catal. B: Environ. **183**, 197–205 (2016)
95. W.Y. Lim, M. Hong, G.W. Ho, In situ photo-assisted deposition and photocatalysis of $ZnIn_2S_4$/transition metal chalcogenides for enhanced degradation and hydrogen evolution under visible light. Dalton Trans. **45**, 552–560 (2016)
96. S. Shen, J. Chen, X. Wang, L. Zhao, L. Guo, Microwave-assisted hydrothermal synthesis of transition-metal doped $ZnIn_2S_4$ and its photocata-lytic activity for hydrogen evolution under visible light. J. Power Sources **196**, 10112–10119 (2011)
97. H. Kaga, A. Kudo, Cosubstituting effects of copper(I) and gallium(III) for $ZnGa_2S_4$ with defect chalcopyrite structure on photocatalytic activity for hydrogen evolution. J. Catal. **310**, 31–36 (2014)
98. T.A. Kandiel, D.H. Anjum, P. Sautet, T.L. Bahersd, K. Takanabe, Electronic structure and photocatalytic activity of wurtzite Cu–Ga–S nano-crystals and their Zn substitution. J. Mater. Chem. A **3**, 8896–8904 (2015)
99. P. Wu, Z. Zhou, J. Shi, L. Guo, First-principles calculations of Cd1−xZnxS doped with alkaline earth metals for photocatalytic hydrogen genera-tion. Int. J. Hydrogen Energy **37**, 13074–13081 (2012)
100. L. Wang, W.Z. Wang, M. Shang, W.Z. Yin, S.M. Sun, L. Zhang, Int. J. Hydrogen Energy **35**, 19 (2010)
101. F. del Valle, A. Ishikawa, K. Domen, J.A.V. de la Mano, M.C. Sanchez-Sanchez, I.D. Gonzalez, S. Herreras, N. Mota, M.E. Rivas, M.C.A. Galvan, J.L.G. Fierro, R.M. Navarro, Catal. Today **143**, 51 (2009)
102. D.L. Chen, L. Gao, Solid State Commun. **133**, 145 (2005)
103. S.Q. Peng, Y.J. Peng, Y.X. Li, G.X. Lu, S.B. Li, Res. Chem. Intermed. **35**, 739 (2009)
104. C.J. Xing, Y.J. Zhang, W. Yan, L.J. Guo, Int. J. Hydrogen Energy **2006**, 31 (2018)
105. J. Zhang, J.G. Yu, M. Jaroniec, J.R. Gong, Noble metal-free reduced graphene oxide-ZnxCd1 − xS Nanocomposite with enhanced solar photocatalytic H_2–production performance. Nano Lett. **12**, 4584–4589 (2012)
106. X. Li, J. Yu, J. Low, Y. Fang, J. Xiao, X. Chen, Engineering hetero-geneous semiconductors for solar water splitting. J. Mater. Chem. A **3**, 2485–2534 (2015)
107. M. Gao, Y. Xu, J. Jiang, S. Yu, Nanostructured metal chalcogenides: synthesis, modification, and applications in energy conversion and storage devices. Chem. Soc. Rev. **42**, 2986–3017 (2013)
108. K. Zhang, L. Guo, Metal sulphide semiconductors for photocatalytic hydrogen production. Catal. Sci. Technol. **3**, 1672–1690 (2013)
109. X. Yu, X. An, A. Shavel, M. Ibanez, A. Cabot, The effect of the Ga con-tent on the photocatalytic hydrogen evolution of CuIn1 − xGaxS2 nanocrystals. J. Mater. Chem. A **2**, 12317–12322 (2014)
110. N. Liang, Q. He, S. Huang, M. Wang, W. Chen, M. Xu, Y. Yuan, J. Zai, N. Fang, X. Qian, AgInxGa1 − xS2 solid solution nanocrystals: synthesis, band gap tuning and photocatalytic activity. Cryst. Eng. Comm. **16**, 10123–10130 (2014)
111. M. Regulacio, M. Han, Multinary I-III-VI2 and I2-II-IV-VI4 semicon-ductor nanostructures for photocatalytic applications. Acc. Chem. Res. **49**, 511–519 (2016)
112. M.V. Kovalenko, L. Manna, A. Cabot, Z. Hens, D.V. Talapin, C.R. Kagan et al., Prospects of nanoscience with nanocrystals. ACS Nano **9**, 1012–1057 (2015)
113. K. Li, B. Chai, T. Peng, J. Maoa, L. Zan, Synthesis of multicomponent sulfide Ag_2ZnSnS_4 as an efficient photocatalyst for H_2 production under vis-ible light irradiation. RSC Adv. **3**, 253–258 (2013)
114. E. Ha, L.Y.S. Lee, H. Man, S.C.E. Tsang, K. Wong, Morphology-controlled synthesis of Au/Cu_2FeSnS_4 core–shell nanostructures for plasmon-enhanced photocatalytic hydrogen generation. ACS Appl. Mater. Interfaces. **7**, 9072–9077 (2015)

115. S.J. Mun, S.-J. Park, Graphitic carbon nitride materials for photocatalytic hydrogen production via water splitting: a short review. Catalysts **9**, 805 (2019). 10.3390/catal9100805.
116. W.J. Ong, L.L. Tan, Y.H. Ng, S.T. Yong, S.P. Chai, Graphitic carbon nitride ($g-C_3N_4$)-based photocatalysts for artificial photosynthesis and environmental remediation: Are we a step closer to achieving sustainability? Chem. Rev. **116**, 7159–7329 (2016)
117. S. Cao, J. Low, J. Yu, M. Jaroniec, Adv. Mater. **27**, 2150–2176 (2015)
118. W. Iqbal, X. Bo Yang, M.R. Zhao, M. Waqas, Y. Gong, J. Zhang, Y. Mao, Controllable synthesis of graphitic carbon nitride nanomaterials for solar energy conversion and environmental remediation: the road travelled and the way forward. Catal. Sci. Technol. **8**, 4576–4599 (2018)
119. A. Naseri, M. Samadi, A. Pourjavadi, A.Z. Moshfegh, S. Ramakrishna, Graphitic carbon nitride ($g-C_3N_4$)-based photocatalysts for solar hydrogen generation: recent advances and future development directions. J. Mater. Chem. A **5**, 23406–23433 (2017)
120. M.S. Nasir, G. Yang, I. Ayub, S. Wang, L. Wang, X. Wang, S. Ramakarishna, Recent development in graphitic carbon nitride based photocatalysis for hydrogen generation. Appl. Catal. B Environ. **257**, 117855 (2019)
121. G. Mamba, A.K. Mishra, Graphitic carbon nitride ($g-C_3N_4$) nanocomposites: a new and exciting generation of visible light driven photocatalysts for environmental pollution remediation. Appl. Catal. B Environ. **198**, 347–377 (2016)
122. G. Liao, S. Chen, X. Quan, H. Yu, H. Zhao, Graphene oxide modified $g-C_3N_4$ hybrid with enhanced photocatalytic capability under visible light irradiation. J. Mater. Chem. **22**, 2721–2726 (2012)
123. S. Hen, C. Wang, B.R. Bunes, Y. Li, C. Wang, L. Zang, Enhancement of visible-light-driven photocatalytic H2 evolution from water over $g-C_3N_4$ through combination with perylene diimide aggregates. Appl. Catal. A Gen. **498**, 63–68 (2015)
124. L. Jiang, X. Yuan, Y. Pan, J. Liang, G. Zeng, Z. Wu, H. Wang, Doping of graphitic carbon nitride for photocatalysis: a reveiw. Appl. Catal. B Environ. **217**, 388–406 (2017)
125. Y.-P. Zhu, T.-Z. Ren, Z.-Y. Yuan, Mesoporous phosphorus-doped $g-C_3N_4$ nanostructured flowers with superior photocatalytic hydrogen evolution performance. ACS Appl. Mater. Interfaces. **7**, 16850–16856 (2015)
126. S. Yang, J. Gong, J. Zhang, L. Zhan, L. Ma, Exfoliated graphitic carbon nitride nanosheets as efficient catalysts for hydrogen evolution under visible light. Adv. Mater. **25**, 2452–2456 (2013)
127. Z. Zhao, X. Wang, Z. Shu, J. Zhou, T. Li, W. Wang, Y. Tan, Facile preparation of hollow-nanosphere based mesoporous $g-C_3N_4$ for highly enhanced visible-light-driven photocatalytic hydrogen evolution. Appl. Surf. Sci. **455**, 591–598 (2018)
128. Y. He, Y. Wang, L. Zhang, B. Teng, M. Fan, High-e_ciency conversion of CO_2 to fuel over $ZnO/g-C_3N_4$ photocatalyst. Appl. Catal. B Environ. **168**, 1–8 (2015)
129. Z. Zhang, K. Liu, Z. Feng, Y. Bao, B. Dong, Hierarchical sheet-on-sheet ZnIn2S4/g-C3N4 heterostructure with highly e_cient photocatalytic H2 production based on photoinduced interfacial charge transfer. Sci. Rep. **6**, 19221 (2016)
130. J. Yuan, J. Wen, Y. Zhong, X. Li, Y. Fang, S. Zhang, W. Liu, Enhanced photocatalytic H2 evolution over noble-metal-free NiS cocatalyst modified CdS nanorods/$g-C_3N_4$ heterojunctions. J. Mater. Chem. A **3**, 18244–18255 (2015)
131. J. Zhang, Z. Ma, Ag6Mo10O33/g-C3N4 1D-2D hybridized heterojunction as an e_cient visible-light-driven photocatalyst. Mol. Catal. **432**, 285–291 (2017)
132. Y. Tan, Z. Shu, J. Zhou, T. Li, W. Wang, Z. Zhao, One-step synthesis of nanostructured $g-C_3N_4/TiO_2$ composite for highly enhanced visible-light photocatalytic H_2 evolution. Appl. Catal. B Environ. **230**, 260–268 (2018)
133. H. Zhang, F. Liu, H. Wu, X. Cao, J. Sun, W. Lei, In situ synthesis of $g-C_3N_4/TiO_2$ heterostructures with enhanced photocatalytic hydrogen evolution under visible light. RSC Adv. **7**, 40327–40333 (2017)
134. Q. Guo, Y. Zhang, J. Qiu, G. Dong, Engineering the electronic structure and optical properties of g-C3N4 by non-metal ion doping. J. Mater. Chem. C **4**, 6839–6847 (2016)

135. J. Feng, D. Zhang, H. Zhou, M. Pi, X. Wang, S. Chen, Coupling P nanostructures with P-doped g-C_3N_4 as e_cient visible light photocatalysts for H_2 evolution and RhB degradation. ACS Sustain. Chem. Eng. **6**, 6342–6349 (2018)
136. Y.H. Hu, A highly efficient photocatalyst—hydrogenated black TiO_2 for the photocatalytic splitting of water. Angew. Chem. Int. Ed. **51**, 12410–12412 (2012)
137. W. Li, Y. Deng, Z. Wu, X. Qian, J. Yang, Y. Wang, D. Gu, F. Zhang, B. Tu, D. Zhao, Hydrothermal etching assisted crystallization: a facile route to functional yolkshell titanate microspheres with ultrathin nanosheets-assembled double shells. J. Am. Chem. Soc. **133**, 15830–15833 (2011)
138. A. Naldoni, M. Altomare, G. Zoppellaro, N. Liu, Š. Kment, R. Zbořil, P. Schmuki, Photocatalysis with reduced TiO_2: from black TiO_2 to cocatalyst-free hydrogen production. ACS Catal. **9**, 345–364 (2019)
139. R. Asahi, T. Morikawa, T. Ohwaki, K. Aoki, Y. Taga, Visible-light photocatalysis in nitrogen-doped titanium oxides. Science **293**, 269 (2001)
140. Y. Gao, K. Qian, B. Xu, Z. Li, J. Zheng, S. Zhao, F. Ding*, Y. Sun, Z. Xu, Recent advances in visible-light-driven conversion of CO_2 by photocatalysts into fuels or value-added chemicals. Carbon Resour. Convers. **3**, 46–59 (2020)
141. D. Li, H. Haneda, S. Hishita, N. Ohashi, Visible-light-driven N − F—codoped TiO_2 Photocatalysts. 2. Optical characterization, photocatalysis, and potential application to air purification, Chem. Mater. **17**, 2596–2602 (2005)
142. J. Schneider, M. Matsuoka, M. Takeuchi, J. Zhang, Y. Horiuchi, M. Anpo, D.W. Bahnemann, Understanding TiO_2 photocatalysis: mechanisms and materials. Chem. Rev. **114**, 9919–9986 (2014)
143. X. Li, Z. Zhuang, W. Li, H. Pan, Photocatalytic reduction of CO_2 over noble metal loaded and nitrogen-doped mesoporous TiO_2. Appl. Catal. A **429–430**, 31–38 (2012)
144. N.R. Khalid, M.K. Hussain, G. Murtaza, M. Ikram, M. Ahmad, A. Hamad, A novel Ag_2O/Fe–TiO_2 photocatalyst for CO_2 conversion into methane under visible light. J. Inorg. Organomet. Polym. Mater. **29**, 288–1296 (2019)
145. O. Ola, M. Mercedes Maroto-Valer, Role of catalyst carriers in CO_2 photoreduction over nanocrystalline nickel loaded TiO_2-based photocatalysts, J. Catal. **309**, 300–308 (2014)
146. K. Koči, K. Matějů, L. Obalova, S. Krejčikova, Z. Lacny, D. Placha, L. Čapek, A. Hospodkova, O. Šolcova, Effect of silver doping on the TiO_2 for photocatalytic reduction of CO_2. Appl. Catal. B **96**, 239–244 (2010)
147. S. Linic, P. Christopher, D.B. Ingram, Plasmonic-metal nanostructures for efficient conversion of solar to chemical energy. Nat. Mater. **10**, 911 (2011)
148. W. Hou, W.H. Hung, P. Pavaskar, A. Goeppert, M. Aykol, S.B. Cronin, Photocatalytic conversion of CO_2 to hydrocarbon fuels via plasmon-enhanced absorption and metallic interband transitions. ACS Catal. **1**, 929–936 (2011)
149. Yu. Bingcheng, Yong Zhou, Peng Li, Tu Wenguang, Ping Li, Lanqin Tang, Jinhua Ye, Zhigang Zou, Photocatalytic reduction of CO_2 over Ag/TiO_2 nanocomposites prepared with a simple and rapid silver mirror method. Nanoscale **8**, 11870–11874 (2016)
150. E. Liu, L. Qi, J. Bian, Y. Chen, X. Hu, J. Fan, H. Liu, C. Zhu, Q. Wang, Afacile strategy to fabricate plasmonic Cu modified TiO_2 nano-flower films for photocatalytic reduction of CO_2 to methanol. Mater. Res. Bull. **68**, 203–209 (2015)
151. M. Cheng, S. Yang, R. Chen, X. Zhu, Q. Liao, Y. Huang, Copper-decorated TiO_2 nanorod thin films in optofluidic planar reactors for e_cient photocatalytic reduction of CO_2. Int. J. Hydrogen Energy **42**, 9722–9732 (2017)
152. Y. Li, W. Zhang, X. Shen, P. Peng, L. Xiong, Y. Yu, Octahedral Cu_2O-modified TiO_2 nanotube arrays for efficient photocatalytic reduction of CO_2. Cuihua Xuebao Chin. J. Catal. **36**, 2229–2236 (2015)
153. X. Li, H. Liu, D. Luo, J. Li, Y. Huang, H. Li, Y. Fang, Y. Xu, L. Zhu, Adsorption of CO_2 on heterostructure CdS(Bi_2S_3)/TiO_2 nanotube photocatalysts and their photocatalytic activities in the reduction of CO_2 to methanol under visible light irradiation. Chem. Eng. J. **180**, 151–158 (2012)

154. G. Williams, B. Seger, P.V. Kamat, ACS Nano **2**, 1487–1491 (2008)
155. B.L. Dasari, J.M. Nouri, D. Brabazon, S. Naher, Energy **140**, 766–778 (2017)
156. J. Wang, X. Jin, C. Li, W. Wang, H. Wu, S. Guo, Chem. Eng. J. **370**, 831–854 (2019)
157. M. Inagaki, F. Kang, J. Mater. Chem. **2**, 13193–13206 (2014)
158. L.-L. Tan, S.-P. Chai, A.R. Mohamed, Chem. Sus. Chem. **5**, 1868–1882 (2012)
159. Y. Zhao, Y. Wei, X. Wu, H. Zheng, Z. Zhao, J. Liu, J. Li, Appl. Catal. B Environ. **226**, 360–372 (2018)
160. S. Sorcar, J. Thompson, Y. Hwang, Y.H. Park, T. Majima, C.A. Grimes, J.R. Durrant, S. Il, Energy Environ. Sci. **11**, 3183–3193 (2018)
161. Y. Rambabua, U. Kumarb, N. Singhalb, M. Kaushalb, M. Jaiswala, S.L. Jainb, S.C. Roy, Photocatalytic reduction of carbon dioxide using graphene oxide wrapped TiO_2 nanotubes. Appl. Surf. Sci. **485**, 48–55 (2019)
162. T.F. Yeh, J.M. Syu, C. Cheng, T.H. Chang, H.S. Teng, Adv. Funct. Mater. 20 (2010)
163. J.-Y. Li, L. Yuan, S.-H. Li, Z.-R. Tang, Y.-J. Xu, One-dimensional copper-based heterostructures toward photo-driven reduction of CO_2 to sustainable fuels and feedstocks. J. Mater. Chem. A **7**, 8676–8689 (2019)
164. H.R. Kim, A. Razzaq, C.A. Grimes, Heterojunction p-n-p Cu_2O/S-TiO_2/CuO: synthesis and application to photocatalytic conversion of CO_2 to methane. J. CO_2 Util. **20**, 91–96 (2017).
165. J.-C. Wang, L. Zhang, W.-X. Fang, J. Ren, Y.-Y. Li, H.-C. Yao, J.-S. Wang, Z.-J. Li, Enhanced photoreduction CO_2 activity over direct Z-scheme α-Fe_2O_3/Cu_2O heterostructures under visible light irradiation. ACS Appl. Mater. Interfaces **7**(16), 8631–8639 (2015)
166. H. Tada, T. Mitsui, T. Kiyonaga, T. Akita, K. Tanaka, Nat. Mater. **5**, 782 (2006)
167. C. Kim, K.M. Cho, A. Al-Saggaf, I. Gereige, H.-T. Jung, ACS Catal. **8**, 4170 (2018)
168. R. Gusain, P. Kumar, O.P. Sharma, S.L. Jain, O.P. Khatri, Reduced graphene oxide–CuO nanocomposites for photocatalytic conversion of CO_2 into methanol under visible light irradiation. Appl. Catal. B **181**, 352–362 (2016)
169. Y. Fang, X. Wang, Photocatalytic CO_2 conversion by polymeric carbon nitrides. Chem. Commun. **54**, 5674–5687 (2018)
170. T.-D. Nguyen, C.-T. Dinh, T.-O. Do, Tailoring the assembly, interfaces, and porosity of nanostructures toward enhanced catalytic activity. Chem. Commun. **51**, 624–635 (2015)
171. G. Zhang, Z.-A. Lan, L. Lin, S. Lin, X. Wang, Overall water splitting by Pt/g-C_3N_4 photocatalysts without using sacrificial agents. Chem. Sci. **7**, 3062–3066 (2016)
172. G. Dong, L. Zhang, Porous structure dependent photoreactivity of graphitic carbon nitride under visible light, J. Mater. Chem. **22**, 1160–1166 (2012)
173. C. Ye, J.-X. Li, Z.-J. Li, X.-B. Li, X.-B. Fan, L.-P. Zhang, B. Chen, C.-H. Tung, L.- Z. Wu, Enhanced driving force and charge separation efficiency of protonated g-C_3N_4 for photocatalytic O_2 evolution. ACS Catal. **5**, 6973–6979 (2015)
174. S. Zhang, J. Li, X. Wang, Y. Huang, M. Zeng, J. Xu, Rationally designed 1D Ag@$AgVO_3$ nanowire/graphene/protonated g-C_3N_4 nanosheet heterojunctions for enhanced photocatalysis via electrostatic self-assembly and photochemical reduction methods. J. Mater. Chem. A **3**, 10119–10126 (2015)
175. K.S. Novoselov, A.K. Geim, S.V. Morozov, D. Jiang, Y. Zhang, S.V. Dubonos, I.V. Grigorieva, A.A. Firsov, Science **84**(306), 666–669 (2004)
176. C.-M. Fung, J.-Y. Tang, L.-L. Tan, A.R. Mohamed, S.-P. Chai, Recent progress in two-dimensional nanomaterials for photocatalytic carbon dioxide transformation into solar fuels. Mater. Today Sustain. xxx (xxxx) xxx
177. G. Shi, L. Yang, Z. Liu, X. Chen, J. Zhou, Y. Yu, Appl. Surf. Sci. **427**, 1165–1173 (2018)
178. W.P.C. Lee, L.L. Tan, S. Sumathi, S.P. Chai, Int. J. Hydrogen Energy **43**, 748–756 (2018)
179. J. Ying Tang, W. Guo Zhou, R. Tang Guo, C. Ying Huang, W. Guo Pan, Catal. Commun. **107**, 92–95 (2018)
180. Z. Sun, S. Wang, Q. Li, M. Lyu, T. Butburee, B. Luo, H. Wang, J.M.T.A. Fischer, C. Zhang, Z. Wu, L. Wang, Adv. Sustain. Syst. **1**, 1700003 (2017)
181. J. Liu, Y. Liu, N. Liu, Y. Han, X. Zhang, H. Huang, Y. Lifshitz, S.-T. Lee, J. Zhong, Z. Kang, Metal-free efficient photocatalyst for stable visible water splitting via a two-electron pathway, Science **347**, 970 (2015).

182. W.-J. Ong, L.-L. Tan, S.-P. Chai, S.-T. Yong, A.R. Mohamed, Surface charge modification via protonation of graphitic carbon nitride (g-C_3N_4) for electrostatic selfassembly construction of 2D/2D reduced graphene oxide (rGO)/g-C_3N_4 nanostructures toward enhanced photocatalytic reduction of carbon dioxide to methane. Nano Energy **13**, 757–770 (2015).
183. W.-J. Ong, L.K. Putri, Y.-C. Tan, L.-L. Tan, N. Li, Y.H. Ng, X. Wen, S.-P. Chai, Unravelling charge carrier dynamics in protonated g-C_3N_4 interfaced with carbon nanodots as co-catalysts toward enhanced photocatalytic CO_2 reduction: a combined experimental and first-principles DFT study. Nano Res. **10**, 1673–1696 (2017)
184. A. Fujishima, K. Honda, Nature **238**, 37–38 (1972)
185. S. Zhang, Y. Zhao, R. Shi, G.I.N. Waterhouse, T. Zhang, Photocatalytic ammonia synthesis: recent progress and future. Energy Chem **1**, 100013 (2019)
186. X. Chen, N. Li, Z. Kong, W.-J. Ong, X. Zhao, Photocatalytic fixation of nitrogen to ammonia:state-of-the-art advancements and future prospects, Mater. Horiz, https://doi.org/10.1039/c7mh00557a
187. S. Bourgeois, D. Diakite, M. Perdereau, React. Solids **6**, 95–104 (1988)
188. H. Hirakawa, M. Hashimoto, Y. Shiraishi, T. Hirai, Photocatalytic conversion of nitrogen to ammonia with water on surface oxygen vacancies of titanium dioxide. J. Am. Chem. Soc. **139**, 10929–10936 (2017)
189. G. Zhang, X. Yang, C. He, P. Zhang, H. Mi, J. Mater. Chem. A (2019). https://doi.org/10.1039/C9TA10471B
190. G.N. Schrauzer, T.D. Guth, J. Am. Chem. Soc. **99**, 7189–7193 (1977)
191. V. Augugliaro, A. Lauricella, L. Rizzuti, M. Schiavello, A. Sclafani, Conversion of solar-energy to chemical energy by photoassisted processes-I. Preliminary result/s on ammonia production over doped titanium dioxide catalysts in a fluidized-bed reactor. Int. J. Hydrog. Energy **7**, 845–849 (1982)
192. M. Ali, F. Zhou, K. Chen, C. Kotzur, C. Xiao, L. Bourgeois, X. Zhang, D.R. MacFarlane, Nanostructured photoelectrochemi- cal solar cell for nitrogen reduction using plasmon-enhanced black silicon. Nat. Commun. **7**, 11335 (2016)
193. T. Oshikiri, K. Ueno, H. Misawa, Plasmon-induced ammonia syn- thesis through nitrogen photofixation w ith v isible light irradiation. Angew. Chem. Int. Ed. **53**, 9802–9805 (2014)
194. H. Zeng, S. Terazono, T. Tanuma, A novel catalyst for ammonia synthesis at ambient temperature and pressure: visible light re- sponsive photocatalyst using localized surface plasmon resonance. Catal. Commun. **59**, 40–44 (2015)
195. J. Yang, Y. Guo, R. Jiang, F. Qin, H. Zhang, W. Lu, J. Wang, J.C. Yu, High-efficiency, "working-in-tandem" nitrogen photofix- ation achieved by assembling plasmonic gold nanocrystals on ultrathin titania nanosheets. J. Am. Chem. Soc. **140**, 8497–8508 (2018)
196. H. Li, J. Shang, Z. Ai, L. Zhang, Efficient visible light ni- trogen fixation with BiOBr nanosheets of oxygen vacancies on the exposed 001 facets. J. Am. Chem. Soc. **137**, 6393–6399 (2015)
197. J. Li, Y. Yu, Zhang, L Bismuth oxyhalide nanomaterials: layered structures meet photocatalysis. Nanoscale **6**, 8473–8488 (2014)
198. X.Y. Kong, W.P.C. Lee, W.-J. Ong, S.-P. Chai, A.R. Mohamed, Oxygen-deficient BioBr as a highly stable photocatalyst for efficient CO_2 reduction into renewable carbon-neutral fuels. Chem. Cat. Chem. **8**, 3074–3081 (2016)
199. S. Wang, X. Hai, X. Ding, K. Chang, Y. Xiang, X. Meng, Z. Yang, H. Chen, J. Ye, Adv. Mater. **29**, 1701774 (2017)
200. S.M. Sun, Q. An, W.Z. Wang, L. Zhang, J.J. Liu, W.A. God- dard, Efficient photocatalytic reduction of dinitrogen to ammonia on bismuth monoxide quantum dots. J. Mater. Chem. A **5**, 201–209 (2017)
201. Y. Hao, X. Dong, S. Zhai, H. Ma, X. Wang, X. Zhang, Hy- drogenated bismuth molybdate nanoframe for efficient sunlight–driven nitrogen fixation from air. Chem. Eur. J. **22**, 18722–18728 (2016)
202. C. Xiao, H. Hu, X. Zhang, D.R. MacFarlane, Nanostructured gold/bismutite hybrid heterocatalysts for plasmon-enhanced pho- tosynthesis of ammonia. ACS Sustain. Chem. Eng. **5**, 10858–10863 (2017)

203. X. Zong, H. Yan, G. Wu, G. Ma, F. Wen, L. Wang, C. Li, En- hancement of photocatalytic H_2 evolution on CdS by loading MoS_2 as cocatalyst under visible light irradiation. J. Am. Chem. Soc. **130**, 7176–7177 (2008)
204. H. Miyama, N. Fujii, Y. Nagae, Heterogeneous photocatalytic synthesis of ammonia from water and nitrogen. Chem. Phys. Lett. **74**, 523–524 (1980)
205. M.M.T. Khan, R.C. Bhardwaj, C. Bhardwaj, Catalytic fixation of nitrogen by the photocatalytic CdS/Pt/RuO_2 particulate system in the presence of aqueous [Ru(Hedta)N_2]-complex. Angew. Chem. Int. Ed. **27**, 923–925 (1988)
206. L. Ye, C. Han, Z. Ma, Y. Leng, J. Li, X. Ji, D. Bi, H. Xie, Z. Huang, Ni 2 P loading on Cd 0.5 Zn 0.5 S solid solution for excep- tional photocatalytic nitrogen fixation under visible light. Chem. Eng. J. **307**, 311–318 (2017)
207. K. A. Brown, D. F. Harris, M. B. Wilker, A. Rasmussen, N. Khadka, H. Hamby, S. Keable, G. Dukovic, J. W. Peters and L. C. Seefeldt, Science, 2016, 352, 448–450
208. A. Banerjee, B.D. Yuhas, E.A. Margulies, Y. Zhang, Y. Shim, M.R. Wasielewski, M.G. Kanatzidis, J. Am. Chem. Soc. **137**, 2030–2034 (2015)
209. S. Hu, X. Chen, Q. Li, Y. Zhao, W. Mao, Catal. Sci. Technol. **6**, 5884–5890 (2016)
210. Q. Zhang, S. Hu, Z. Fan, D. Liu, Y. Zhao, H. Ma, F. Li, Dalton Trans. **45**, 3497–3505 (2016)
211. S. Sun, X. Li, W. Wang, L. Zhang and X. Sun, Appl. Catal. B. **200**, 323–329 (2017)
212. C.H. Lui, A.J. Frenzel, D.V. Pilon, Y.H. Lee, X. Ling, G.M. Akselrod, J. Kong, N. Gedik, Phys. Rev. Lett. **113**, 166801 (2014)
213. D. Zhu, L. Zhang, R.E. Ruther, R.J. Hamers, Nat. Mater. **12**, 836–841 (2013)
214. J.A. Bandy, D. Zhu, R.J. Hamers, Diamond Relat. Mater. **64**, 34–41 (2016)
215. G. Dong, W. Ho, C. Wang, J. Mater. Chem. A **3**, 23435–23441 (2015)
216. Y. Shiraishi, S. Shiota, Y. Kofuji, M. Hashimoto, K. Chishiro, H. Hirakawa et al., Nitrogen fixation with water on carbon-nitride-based metal-free photocatalysts with 0.1% solar-to-ammonia energy conversion efficiency. ACS Appl. Energy. Mater. **1**, 4169e77 (2018)
217. Y. Xue, X. Kong, Y. Guo, Z. Liang, H. Cui, J. Tian, Synthesis of porous few-layer carbon nitride with excellent photocatalytic nitrogen fixation. J. Materiomics **6**, 128–137 (2020)
218. E. Vesali-Kermani, A. Habibi-Yangjeh, H. Diarmand-Khalilabad, S. Ghosh, Nitrogen photofixation ability of g-C_3N_4 nanosheets/Bi_2MoO_6 heterojunction photocatalyst under visible-light illumination. J. Colloid Interface Sci. **563**, 81–91 (2020)
219. S. Hu, X. Chen, Q. Li, F. Li, Z. Fan, H. Wang, Y. Wang, B. Zheng and G. Wu, Appl. Catal. B. **201**, 58–69 (2017)
220. Z. Ma, S. Zhao, X. Xiong, B. Hu, C. Song, Catal. Lett. **146**, 2324–2329 (2016)
221. H. Ma, Z. Shi, Q. Li, S. Li, J. Phys. Chem. Solids **99**, 51–58 (2016)
222. H. Diarmand-Khalilabad, A. Habibi-Yangjeh, D. Seifzadeh, S. Asadzadeh-Khaneghah, E. Vesali-Kermani, g-C_3N_4 nanosheets decorated with carbon dots and CdS nanoparticles: Novel nanocomposites with excellent nitrogen photofixation ability under simulated solar irradiation. Ceram. Int. **45**, 2542–2555 (2019)

Chapter 5
Nanomaterials for Photocatalytic Environmental Remediation

5.1 Nanomaterials for Air Purification

The main categories of nanomaterials used for photocatalytic air purification include metal oxides and complex metal oxides = based nanomaterials [1–3]. The photocatalytic performances of these nanomaterials greatly influence by material properties such as crystalline structure, morphological, and optical and electronic properties. Therefore, in this section the influence of modification strategies on the performances of abovementioned photocatalysts for air purification will be reviewed.

5.1.1 Metal Oxides-Based Nanomaterials

5.1.1.1 TiO_2

TiO_2 is a broadly studied nanomaterial because of its extraordinary features including low-cost, nontoxic nature, higher stability, and highly effective photocatalytic efficiency under UV light irradiation [4, 5]. TiO_2 has three crystallographic phases containing anatase, rutile, and brookite which showed variable photocatalytic activity. Among these, anatase phase has exhibited higher photocatalytic efficiencies because of lesser reunion of charges and high ratio of hydroxyl radicals [6].

It was found that TiO_2 alteration is a good method in order to get excellent photocatalysis performance and good stability in air purification. Currently, transition metal doping, semiconductor coupling, loading with noble metal, and anion doping have been applied to increase photocatalysis efficiency in air purification [7–10]. For example, Pham et al. [11] studied V-doped TiO_2 with improved photocatalysis efficiency. Authors incorporated V/TiO_2 material into polyurethane having high porosity, followed by applying in a photocatalytic reactor for toluene degradation. Qui et al. [12] deposited Cu_xO onto TiO_2 to get a low-risk reducing nanomaterial in

M. B. Tahir and K. N. Riaz, *Nanomaterials and Photocatalysis in Chemistry*,
Materials Horizons: From Nature to Nanomaterials,
https://doi.org/10.1007/978-981-16-0646-5_5

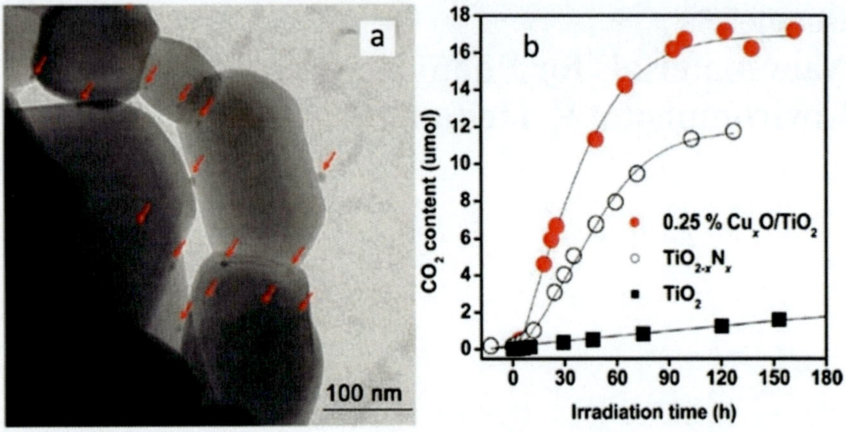

Fig. 5.1 **a** TEM image of the 0.25% Cu_xO-TiO_2 photocatalyst. Clusters of CuxO (showed by red arrows) were highly deposited on surface of TiO_2; **b** Comparative study of CO_2 production using pristine TiO_2, $TiO_{2x}N_x$, and 0.25% CuxO-TiO_2 photocatalyst under the similar given environments [12]. (reproduced from [12], with permission from ACS, 2019)

indoor environment application (Fig. 5.1a). The Cu^{II} in copper oxide favors titanium dioxide during photocatalysis oxidation of volatile organic compounds when exposed to visible photo-illumination. Thus, authors observed good removal of VOCs and anti-pathogenic performance with Cu_xO/TiO_2 composites having optimum concentration of Cu^I and Cu^{II} in Cu_xO. Moreover, Cu_xO-TiO_2 composite showed a good photocatalysis efficiency compared to $TiO_{2-x}N_x$ and pure TiO_2 photocatalysts as shown in Fig. 5.1b. In another report, Jo et al. [13] designed a container layered with pristine TiO_2 or N/TiO_2 with superior removal rate for VOCs in inside atmosphere. The authors found that the photocatalysis with N-doped TiO_2 was greatly effective compared to pristine TiO_2 photocatalyst because it removed about 90% of targeted pollutants under given conditions.

The use of carbon nanomaterials modified TiO_2 has also demonstrated improved photocatalysis efficiency in NO_x removal. It was found that by loading TiO_2 with graphene, the efficiency as photocatalyst increased to about 31–43% in TiO_2-GO and TiO_2-rGO as compared to pure TiO_2 [14, 15]. The enhanced performance was credited to good separation of charges due to graphene incorporation. Recently, Li et al. [16] prepared a Pt-rGO-TiO_2 nanomaterial (Fig. 5.2a) with a wide range capability of light absorption (800–2500 nm). The novel hybrid photocatalyst exhibited higher performance for decomposition of VOCs under infrared illumination. As presented in Fig. 5.2b, light intensity can significantly influence Pt-rGO-TiO_2 performance on toluene reduction and CO_2 formation. It can be seen that maximum photo-oxidation performance (95%) for toluene was got when infrared light intensity was 116 mW/cm^2.

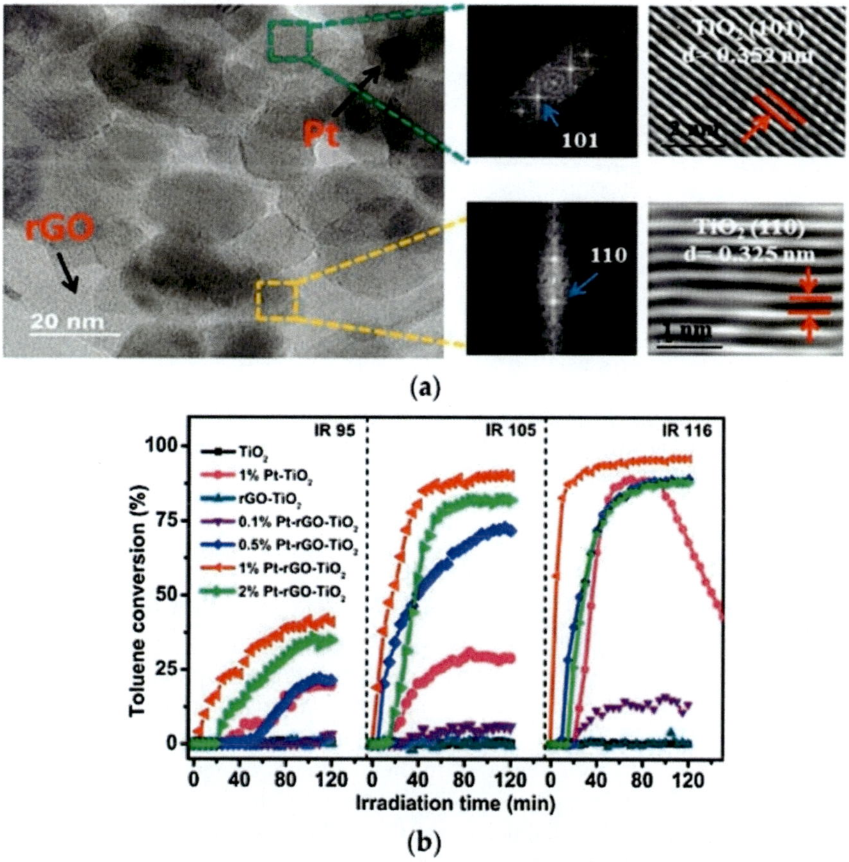

Fig. 5.2 Higher angle annular dark-scanning TEM images and (**b**) time-dependent toluene conversion using pristine TiO_2, 1% Pt-TiO_2, and x% Pt-rGO-TiO_2 (x = 0, 0.1, 0.5, 1, and 2) under infrared illumination with different light intensities [16]. (reproduced from [16], with permission from Elsevier, 2019)

5.1.1.2 Other Metal Oxides

ZnO is the second most investigated and well-known photocatalyst used for photocatalytic air purification applications [17–20]. Nagaragu et al. [19] have reported Ag-ZnO, Cd-ZnO, and Pb-ZnO photocatalytic materials. They observed that Pb/ZnO was highly influencing in decomposition of chlorobenzene when different light sources were applied. Moreover, Kowsari et al. [20] studied a heterojunction based on ZnO/CdS photocatalyst for disintegration of air contaminants. The presence of cadmium oxide nanoparticles on heterostructure surface resulted into higher active surface area and nanorods-like morphology of ZnO, which favors rapid transfer of electrons and superior photocatalysis activity. Figure 5.3a displays the transportation

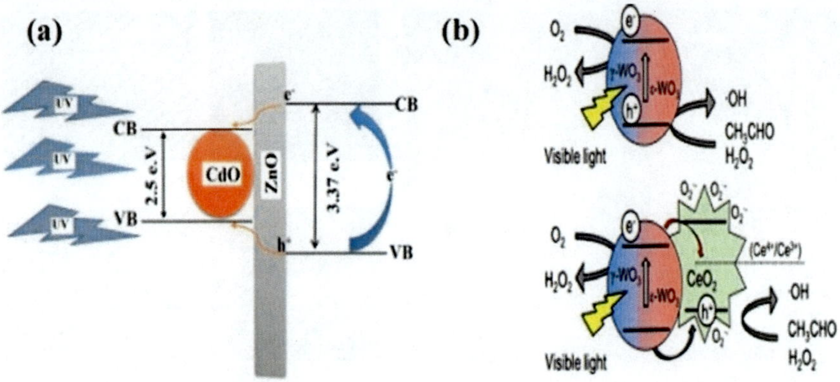

Fig. 5.3 **a** Schematically illustrated of charge carrier's separation and their transport between ZnO and CdO under ultraviolet light, Reproduced from Ref. [20], with permission from Elsevier, copyright (2016), **b** scheme of proposed photocatalysis mechanism for (**a**) plasma-prepared pristine WO_3 (g/e-WO_3) and (b) WO_3/CeO_2 photocatalyst, Reproduced from Ref. [21], with permission from Elsevier, copyright (2017)

as well as separation mechanism of photocreated charge carriers between ZnO and CdO when ultraviolet light was used as an illuminator.

Tungsten trioxide (WO_3) with band gap of 2.8 eV is a superior choice as visible photo-active photocatalytic material for air pollutants degradation [1]. However, due to low energy of conduction band electrons, it has no good capacity to reduce oxygen properly which becomes the source of recombination of charge carriers, followed by a decrease in photocatalytic activity. Therefore, in order to resolve this issue, Wicaksana et al. [22] deposited noble metal Pt on WO_3 which exhibited enhanced performance for acetaldehyde photodegradation. Jansson et al. prepared visible-light active zeolite-WO_3–Pt nanohybrid photocatalyst [23]. They measured their photocatalytic properties under different conditions in the removal of different air pollutants. This novel hybrid system demonstrated superior photocatalytic activity for trichloroethylene degradation when irradiated to visible-light inside reactor.

In another report, plasma-assisted preparation of WO_3 nanoparticles with good structure and excellent capability to lower charge carrier's recombination was reported by Fukumura et al. [21]. Moreover, the colloidal CeO_2 was applied as binder for tungsten trioxide nanoparticles to develop thin films on simple glass substrate. The resulting material displayed higher efficiencies during the photocatalysis removal of acetaldehyde under visible photo-illumination. As drawn in schematic diagram of Fig. 5.3b, synergistic effects between these two materials were the main source for higher photocatalysis activity. The authors proposed that there was a possible reason that cerium dioxide that CeO_2 generated superoxide ($O_2^{-\cdot}$) anions via reversible transition between Ce^{4+} and Ce^{3+} by capturing photocreated electrons generated from WO_3, resulting to lower reunion of charges and higher photocatalysis performance.

Hematite (α-Fe_2O_3) is an appealing n-type semiconducting nanomaterial and widely applied in photocatalysis field because it is most stable phase of Fe_2O_3.

Its band gap value varies from 1.90 to 2.2 eV [24]. The α-Fe_2O_3 has good properties of surface structure such as surface area, porosity, and morphology which is greatly a stimulus photocatalysis performance of nanomaterial. Recently, Balbuena et al. [25] prepared highly porous architecture of α-Fe_2O_3 nanotubes or nanoflakes. Due to good porosity (nanotubes or flakes), the prepared nanomaterial exhibited good photocatalytic performance for NO removal which is comparable to commercial TiO_2 or other highly efficient nanomaterials. Moreover, α-Fe_2O_3 porous architectures displayed better selectivity for NO_x compared to P25.

5.1.1.3 Complex Metal Oxides

Bismuth vanadate ($BiVO_4$) has been considered as an outstanding nanomaterial in photocatalysis field [2]. It has three crystallographic forms such as monoclinic clinobisvanite, orthorhombic pucherite, and tetragonaldreyerite. The monoclinic structure has displayed higher photocatalytic efficiency compared to other phases because it has advantages of small band, low cost, and higher stability in aqueous medium [26, 27]. However, major problem with material is low rate of charge carrier mobility, which results into higher recombination of electron–hole pairs during photocatalysis [26]. Recently, Yang et al. [28] prepared an efficient composite photocatalyst based on rGO-aerogel loaded $BiVO_4$ quantum dots. In the composite system, rGO provided rapid transport path for photocreated electrons, thus increased separation capability of material for charges and improved photocatalysis performance for formaldehyde removal was achieved. The $BiVO_4$ quantum dots possess much higher conduction band than bulk $BiVO_4$ due to quantum size effect, resulting into lower reunion of charges and higher photocatalysis performance.

Recently, researchers have focused on perovskite nanomaterials, which demonstrated good photocatalytic efficiency. The promising perovskites with superior photocatalytic performance are $CaTiO_3$, $BaTiO_3$, $LaFeO_3$, and $LaNiO_3$ [29–31]. The perovskite such as $SrTiO_3$ has showed much higher ability and found rapid implementation in air purification [32, 33].

5.2 Nanomaterials for Wastewater Remediation

The development of nanomaterials for photocatalytic applications has attained a worldwide interest for decades. Nanomaterials provide a great potential for wastewater remediation with high performance which ultimately lead to an increase of water supply through various water sources. In this section, nanomaterials such as metal oxide (TiO_2, ZnO, CeO_2, Fe_2O_3, CuO) and metal sulfide (ZnS, CdS, Ag_2S, Bi_2S_3, CuS) used in removal of wastewater contaminants will be summarized.

5.2.1 *Titanium Dioxide-Based Nanomaterials*

The most efficient and extensively applied nanomaterial in all applications of photocatalysis is TiO_2. It is cheap, chemically stable, and nontoxic materials and thus best choice for photocatalysis applications [34, 35]. The photocatalysis becomes most economical and facile technique for environmental remediation when TiO_2 is used as photocatalyst material. A wide range of wastewater pollutants was degraded when TiO_2 photocatalyst was exposed to ultraviolet light [36, 37]. Therefore, noteworthy progress was done to advance and modify TiO_2-based nanomaterials as an active photocatalyst under visible light.

5.2.1.1 Cation-Doped TiO_2

The transition metal ions including Pd^{2+}, Cu^{2+}, Cd^{2+}, Cr^{2+}, V^{5+}, Fe^{3+}, and Ni^{2+} were incorporated to advance crystalline structure of TiO_2 via creating a donor and/or acceptor levels inside band gap of material. This technique greatly supported the low energy photo-excitation of the catalyst material and revealed a greater photocatalysis efficiency when exposed to photo-illumination [38]. Secondly, dopant can also overcome reunion of charges, resulting to higher quantum activity of TiO_2 [39, 40]. Moreover, it was observed by Karakitsou et al. [41] that modifying TiO_2 with transition metal ions with a valency higher than +4 could increase the photocatalysis performance. Some researchers also found that TiO_2 alteration with trivalent or pentavalent impurity of metal ions can reduce the photocatalysis performance due to acting as trapping centers for photocreated charges [42]. Overall, transition metal ions were found effective due to their two or more oxidation states which supports to enhance TiO_2 photocatalysis performance [43]. For example, Fe_2O_3 has various oxidation states such as Fe^{4+}, Fe^{3+}, and Fe^{2+} which acts as electron capture centers for charge carriers and thereby decrease recombination of electron–hole pairs [44]. Furthermore, the ionic radius of Fe^{3+} (0.79 Å) is comparable to Ti^{4+} (0.75 Å), which favors the easy substitution of Fe^{3+} into the crystal lattice of TiO_2 [45, 46]. Sood et al. [47] prepared Fe^{3+}-doped TiO_2 particles with smaller size, higher visible-light activity, and good separation ability for charges. They observed higher remediation rate of 92% in 5 h for para-nitrophenol without adding oxidizing reagents. The transition of Fe^{3+} into Fe^{2+} and Fe^{4+} and a large number of produced radicals with higher redox potentials are created which oxidize the organic molecules into intermediate and finally into complete removal of para-nitrophenol and MB dye (Fig. 5.4). No doubt that the metallic cations have displayed positive effects on photocatalysis activity of titanium dioxide. However, it was observed that metal-doped materials suffer from thermal instability and also boost the recombination of photocreated charge carriers [48]. Therefore, other strategies were also developed to inhibit photogenerated electrons and holes reunion and enhance light absorption of TiO_2.

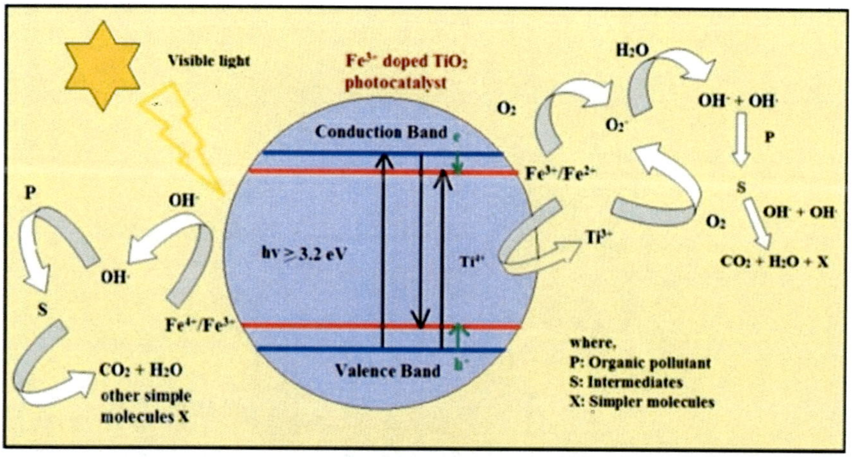

Fig. 5.4 Schematic illustration of mechanism of photocatalysis process occurring over Fe–TiO_2 nanomaterials, reproduced from Ref. [47], with permission from Elsevier, copyright (2015)

5.2.1.2 Anion-Doped TiO_2

Modifying TiO_2 with anionic nonmetals including carbon (C), nitrogen (N), sulfur (S), and iodine (I) has attained much attention [49–56]. Tan et al. [57] found that nonmetal dopants are suitable to increase the morphological and the photocatalysis efficiency of TiO_2. In another example, Ren et al. [58] proved that the nonmetal doping into TiO_2 can decrease its band gap energy and shift its optical absorption in visible-light range. Previous investigations of scientists showed that TiO_2 loaded with anionic nonmetals (carbon, sulfur, fluorine, nitrogen, etc.) demonstrated good effects for visible-light absorption and superior photocatalysis efficiency because impurity levels created by nonmetals in TiO_2 band gap do not act as charge carrier's recombination centers [49, 58–60]. The TiO_2 doping with carbon or nitrogen anions was found much effective for higher photocatalysis activity under visible-light irradiation than other nonmetals [61]. In a study, Saien et al. [62] synthesized a novel N/TiO_2 hybrid modified with hematoporphyrin which showed higher photocatalysis efficiency for the remediation of methyl orange under visible-light illumination. The authors found that nitrogen doping created intra-band states nearer the valence band edge of titanium dioxide, leading to rapid transport of electrons (Fig. 5.5a), and thus, higher visible-light active photocatalytic activity was observed. Similarly, Zhang et al. [63] studied carbon/TiO_2 which demonstrated higher photocatalysis activity in the decomposition of toluene as compared to pure TiO_2. The authors proposed three possible ways which are attributed in increasing the photocatalysis performance of C-doped TiO_2. Firstly, the substitution of carbon at oxygen sites in crystal structure of TiO_2 decreased the band gap energy by overlapping C 2p orbitals with O 2p. Secondly, the carbonaceous material on TiO_2 surface acted as photosensitizer to harvest visible-light and injected electrons into the conduction band of titanium

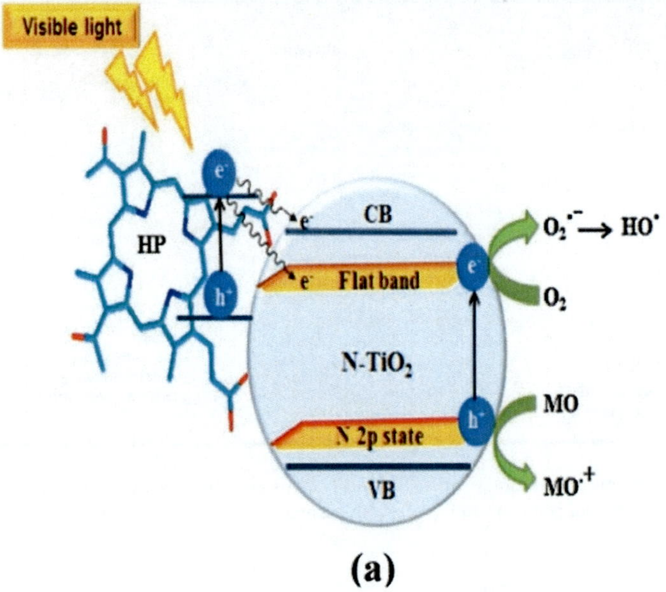

Fig. 5.5 **a** Schematic diagram of proposed reaction of photocatalysis employing N-TiO_2 modified by HP, reproduced from Ref. [62] with permission from ACS, copyright (2007)

dioxide. Thirdly, carbonaceous species also helped to improve separation of charge carriers, and thus, higher visible-light photocatalysis performance was achieved.

5.2.1.3 Noble Metal-Loaded TiO_2

Previously, it was found that nanoparticles of noble metals can capture photo-excited electrons and play a role of sink for interfacial charge transportation. In a report, Tada et al. in their review emphasis on development and application of noble metal nanoparticles incorporated TiO_2 for superior performance in photocatalytic reactions [64]. It was also confirmed that photocatalytic efficiencies of Au, Ag, and Pt nanoparticles had been improved when they were incorporated on the surface of semiconductor nanomaterials [65–67]. Chen et al. [68] studied two-dimensional TiO_2 nanosheet films loaded by Au nanoparticles. The results demonstrated that localized surface plasmonic resonance (LSPR) of Au nanoparticles is highly effective to enhance visible-light absorption properties of Au/TiO_2 photocatalyst system, resulting in superior photocatalysis performance in disintegration of MB contaminant.

The authors observed that upon ultraviolet light irradiation, developed Schottky junction between Au nanoparticles and TiO_2 facilitated the charge separation as

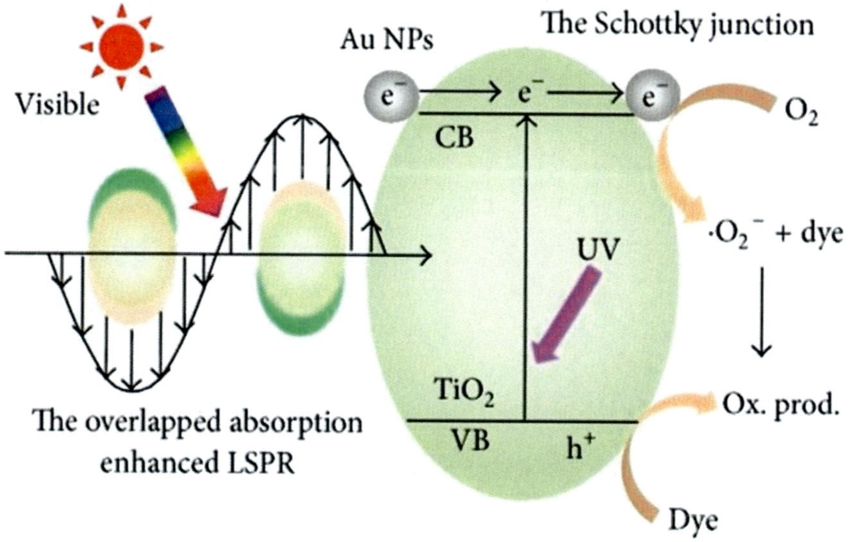

Fig. 5.6 Schematically shown photocatalysis decomposition of methylene blue using Au-TiO_2 nanocomposite with overlapped light capturing nanostructure [68]

described in Fig. 5.6. On the other hand, due to visible-light illumination, the LSPR-created electrons in Au nanoparticles were easily transferred from Au nanoparticles toward TiO_2; therefore, higher photocatalysis activity of Au/TiO_2 system was achieved.

5.2.1.4 Semiconductor Coupled TiO_2

TiO_2 coupling with other semiconductors with varying band gap energy is a very effective technique to decrease reunion of photocreated charges for enhanced photocatalysis performance. A good match between their valence and conduction bands can support a good transport of photo-excited charges from one semiconducting material to another and is displayed in Fig. 5.7a [69]. The difference of energy between band potentials of corresponding bands favors the charge carriers transfer from one semiconductor to other to develop a good separation between holes and electrons. For instance, coupling ZnO with TiO_2 for higher photocatalysis activity is an appropriate method because of their matching band gap energies [70]. Zheng et al. [71] synthesized well-oriented ZnO/TiO_2 nanocomposite to apply for mineralization of methyl orange (MO) dye. The photocatalysis reaction mechanism for MO decomposition was also explained as presented in Fig. 5.7b. It could be perceived that photo-induced electrons were easily transported from CB of ZnO to CB of titanium dioxide under

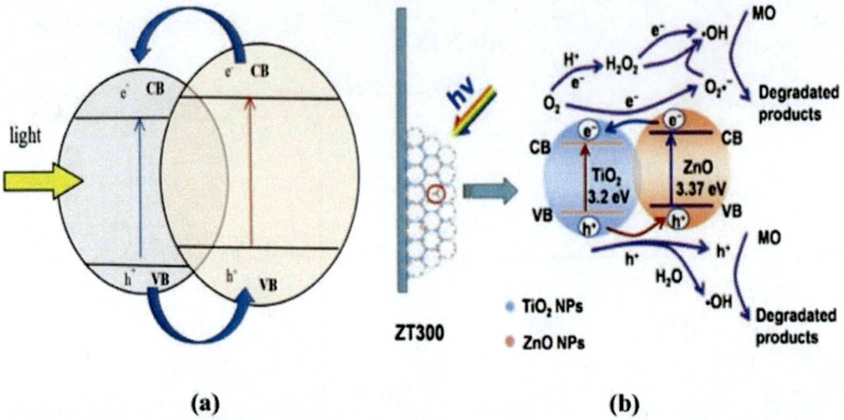

Fig. 5.7 **a** Charge carrier separation with coupled two semiconductors, reproduced from Ref. [69] with permission from Elsevier, copyright (2019), **b** Schematical representation of charge transport phenomenon in the ZT300 composite when exposed to visible-light illumination reproduced from Ref. [71] with permission from Elsevier, copyright (2015)

photo-irradiation and at the same time photocreated positive charge carriers transported from TiO_2 valence band to ZnO valence band, resulting in limited charges reunion.

In another report, Bai et al. [72] determined the photocatalysis efficiency using visible-light illuminated WO_3/TiO_2 heterostructure and observed that photogenerated electrons were transported from TiO_2 conduction band to WO_3 conduction band, and simultaneously, positive charge carriers are transferred in opposite direction because conduction band is lower in potential as compared to TiO_2; this feature was observed to be efficient in reducing the recombination of photoproduced charge species.

5.2.1.5 Graphene-Modified TiO_2

The graphene and its derivatives were used to prepare grapheme-modified TiO_2 nanocomposites due to excellent properties of graphene [73]. Graphene has the ability to separate charge carriers via accepting photocreated charge carriers from host semiconductor material (TiO_2) [74]. Therefore, graphene-TiO_2-based composites were fruitfully used for photocatalysis to remediate water pollutants. Great efforts were made to address the widespread polluted environment because water pollutants come from urban environment and industries [75]. Therefore, graphene-TiO_2-based nanocomposites were used to remediate toxic wastewater pollutants extensively. Chen et al. [76] illustrated a promising mechanism to prepare graphene oxide-TiO_2 composites with varying semiconducting conductivity of graphene oxide (GO). It was seen that graphene optimum contents were useful in photo-electric and photocatalytic efficiency of nanocomposites for methyl orange photodegradation. Furthermore, graphene oxide on TiO_2 surface could also play role of photosensitizer for increased visible-light active photocatalysis efficiency of GO-TiO_2 nanocomposite. In another report, Zhang et al. [77] fabricated TiO_2 (P-25)-graphene nanocomposite photocatalysts via chemical process with enhanced activity. Similarly, many other studies were also conducted to effectively utilize the outstanding properties of graphene in order to increase photocatalytic performance of TiO_2 for degradation of organic pollutants [78–85].

The graphene has showed excellent photocatalytic activity due to the development of C-Ti chemical bonds with TiO_2 which favors the interfacial charge transfer process [86]. Loading Cu(II) on a TiO_2-RGO nanocomposite exhibited about 3 times higher photodegradation rate in phenol decomposition under UV illumination. The authors attributed the higher performance to excellent charge transfer process between graphene and TiO_2 [87]. The efficiency of TiO_2-graphene composite is highly dependent on graphene layers' thickness because increasing thickness decreases the photocatalytic activity [88]. Sakai et al. prepared films of TiO_2-graphene which showed an advanced photogenerated hydrophilic reduction rate of 2.8 times compared to pristine TiO_2 nanosheets because graphene layer functioned as an electron transport mediator [89].

Besides this good progress and positive role of graphene as an electron accepter and transporter, graphene has also capability to form three-dimensional nanostructures via various connections such as H_2 bonding, electrostatic force, and π–π coupling of electrons. Liu et al. synthesized a TiO_2-graphene hydrogel (TGH) through hydrothermal method as displayed in Fig. 5.8a–c [90]. The three-dimensional nanostructure of TGH showed enhanced adsorption capability including physical adsorption and aromatic dye's adsorption on graphene surface through π–π electron coupling due to high surface area and synergistic effects between graphene and TiO_2. Moreover, 3D graphene-TiO2-based system transferred electrons effectively, thus leading to a superior photocatalytic performance for methylene blue degradation (Fig. 5.8d).

Recently, Giovannetti et al. demonstrated the important effects of graphene in graphene-TiO_2 composites applied for photocatalysis removal of synthetic pollutants (dyes) [91]. As shown in Fig. 5.9, in combination with TiO_2 nanostructured, graphene could play a role of co-adsorbent to increase the adsorption on photocatalyst surface due to π-π interactions with dye and as a photosensitizer to transfer electrons from the graphene excited state to the TiO_2 surface with extended visible-light absorption in higher wavelength region of solar spectrum. Furthermore, both functions of graphene as conductive material and co-catalyst support the transport of excited electrons, thus reducing the charge carrier's recombination in the photocatalysis. Additionally, TiO_2 band gap decrease could occur because of contact between graphene and TiO_2 for more visible-light absorption.

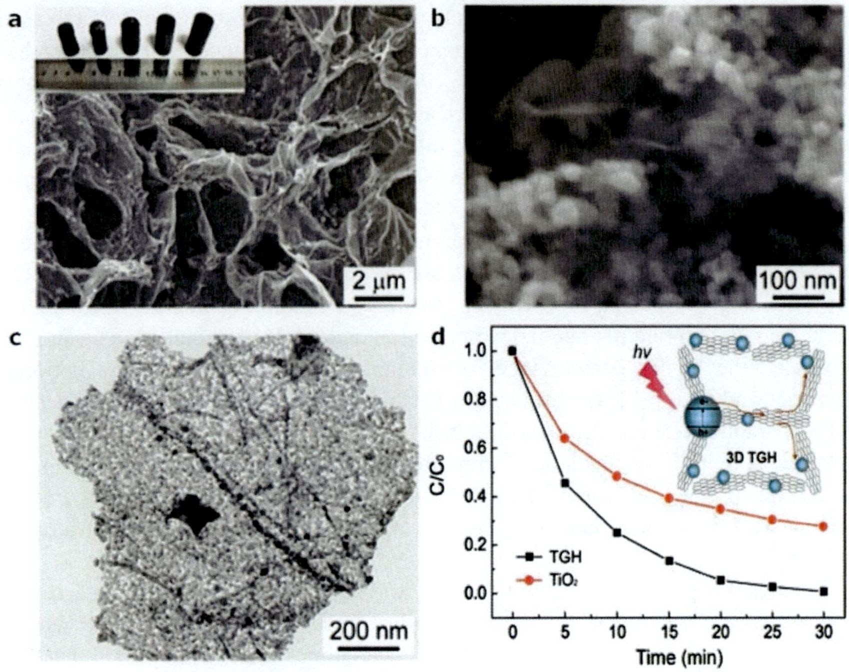

Fig. 5.8 Lower and higher magnification SEM images of TGH (**a**, **b**), inset of (**a**) are the image of TGH with varying ratio of TiO_2 to graphene, (**c**) TGH low-magnified TEM images, (**d**) relative changes of the adsorption peak intensity versus illumination time using TGH and TiO_2 nanomaterials. Inset is the schematically illustrated diagram for charge transport process at interfaces in 3D TGH networks, reproduced from [90] with permission from the ACS, Copyright (2013)

5.2.2 *Zinc Oxide-Based Nanomaterials*

Zinc oxide (ZnO) nanomaterials as photocatalysts have demonstrated great potential due to their higher performance, good aqueous stability, and unexpensive and environment friendly [92]. ZnO has band gap energy (3.37 eV) similar to TiO_2 (3–3.25 eV) and has been applied in the photocatalysis field for wastewater detoxification [93]. The photocatalysis performance of ZnO was compared with other semiconductor photocatalysts for decomposition of azo dyes when solar light illumination was exposed [94]. The authors found that ZnO as a photocatalyst worked well than other nanomaterials being reported, and higher efficiency was ascribed to the capability of ZnO for absorbing a greater part of solar light. The one major problem with ZnO is photocorrosion which gradually decreases the photocatalytic efficiency and limits its potential for long-term stability in photocatalytic applications. Secondly, the rapid recombination of photocreated charge carriers also decreases the photocatalytic performance of ZnO [95]. Recently, different efforts were made to design ZnO-based nanomaterials for higher efficiencies under visible-light radiation.

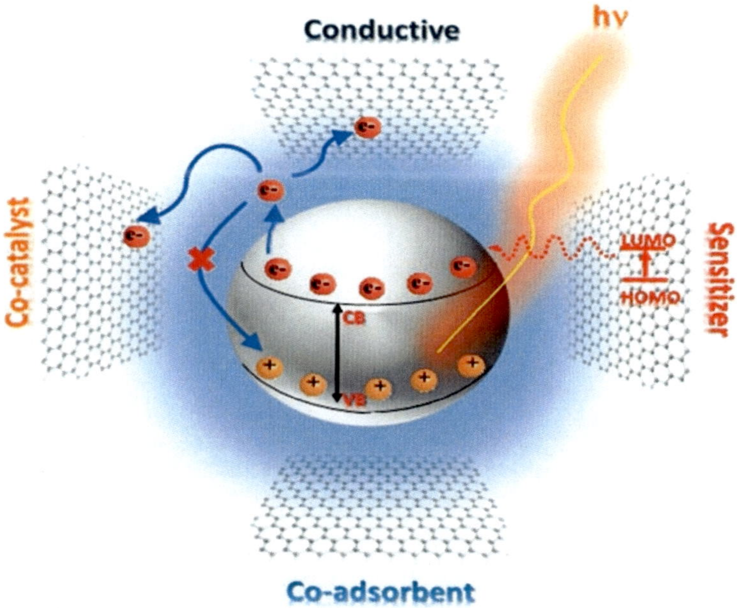

Fig. 5.9 Schematical illustration of main characteristics of graphene in photocatalysis with TiO_2 [91]

Different reports demonstrated that the photocatalysis efficiency of ZnO could be increased under visible-light illumination via doping with rare earth metals including Nd^{3+}, Ln^{3+} [96, 97], deposition with noble metals as Ag, Pd, Pt, Au [98–101] and loading ZnO with other nanomaterials [102, 103]. The rare earth metals have become the main domain of interest in photocatalysis because they have greater ability to capture the photocreated electrons which efficiently overcome recombination of electron–hole pairs. For example, Vaiano et al. [104] have doped ZnO with praseodymium and observed superior photocatalysis efficiency under both UV and visible photo-illumination. Recently, Alam et al. [105] synthesized La, Nd, Sm, and Dy incorporated ZnO nanostructures followed by investigation of their photocatalysis activity for mineralization of organic dyes under ultraviolet light illumination. All metal ions-doped ZnO exhibited enhanced performance toward removal MB dye, of which Nd/ZnO displayed superior efficiency with 98% photocatalytic degradation. The enhanced photocatalytic efficiency of Nd/ZnO was ascribed to well separation of photogenerated charge carriers (Fig. 5.10a), and secondly rare earth metals especially Nd acted as electron traps (Fig. 5.10b) and separated the charge carriers for improved photocatalytic efficiency.

The development of heterojunction/nanocomposite-based photocatalysts for wastewater detoxification is a versatile technique to achieve higher efficiencies. In this regard, modification of ZnO by graphene and graphene-based derivatives shows an important role in increase of photocatalysis activity [106]. Graphene nanosheets

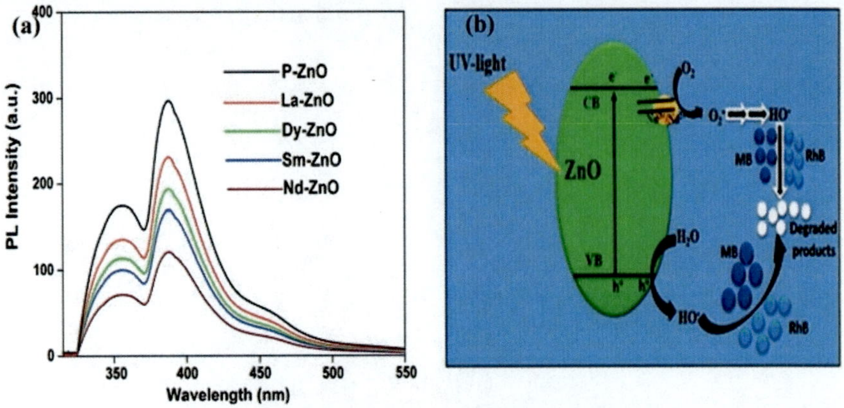

Fig. 5.10 **a** PL spectra of P-ZnO, La-ZnO, Dy-ZnO, Sm-ZnO, and Nd-ZnO at photo-excitation of 320 nm wavelength, **b** schematical representation of plausible reaction mechanism for the removal of MB/RhB over Nd-ZnO under ultraviolet light illumination [105]

in ZnO composites contribute effectively in trapping excited electrons of ZnO and inhibiting electron–hole pairs recombination. A proposed photocatalytic mechanism of wastewater pollutants decomposition using ZnO-graphene composite is displayed in Fig. 5.11 [107]. The photo-excited electrons get reacted with the oxygen (O_2) to create super oxide anionic radicals (O_2^-), and holes produce hydroxyl radicals by

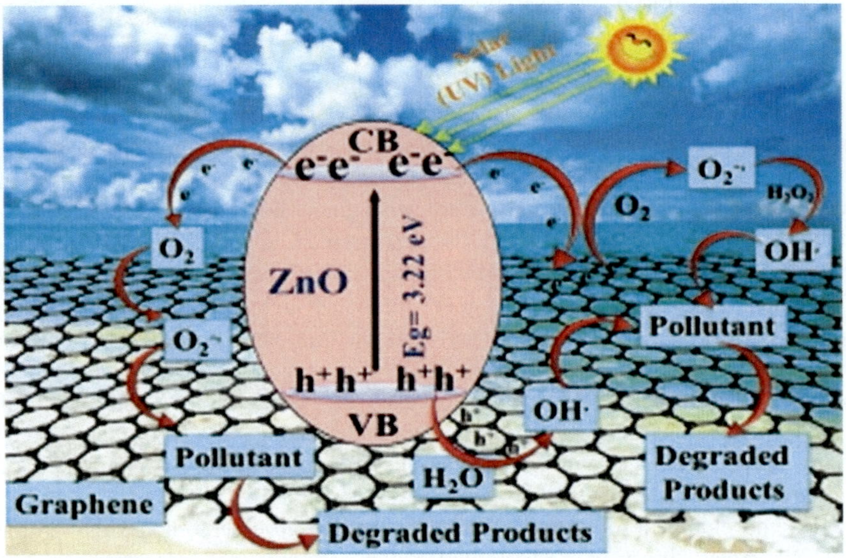

Fig. 5.11 Proposed photocatalysis mechanism of ZnO/grapheme-based nanocomposite for pollutant eradication [107]

reacting with water molecules. These photogenerated radicals react with contaminants adsorbed on surface of ZnO/graphene nanocomposite to degrade the pollutants into harmless molecules. The effective electron transportation due to good electron accepting properties of graphene enhances the photocatalytic activity in ZnO-graphene photocatalytic system [108, 109].

Posa et al. [110] synthesized graphene-ZnO nanostructure nanocomposite via wet chemical process for effective photocatalytic removal of MO dye. Atchudan et al. synthesized ZnO loaded graphene oxide (ZnO/GO) nanocomposite by facile solvothermal technique where ZnO and GO sheet were prepared using thermal oxidation and Hummers technique, respectively (Fig. 5.12a) [111]. The ZnO/GO nanocomposite photocatalyst demonstrated higher degradation performance of 98.5% under UV light irradiation of 15 min compared to pure ZnO (49%) using ultraviolet light

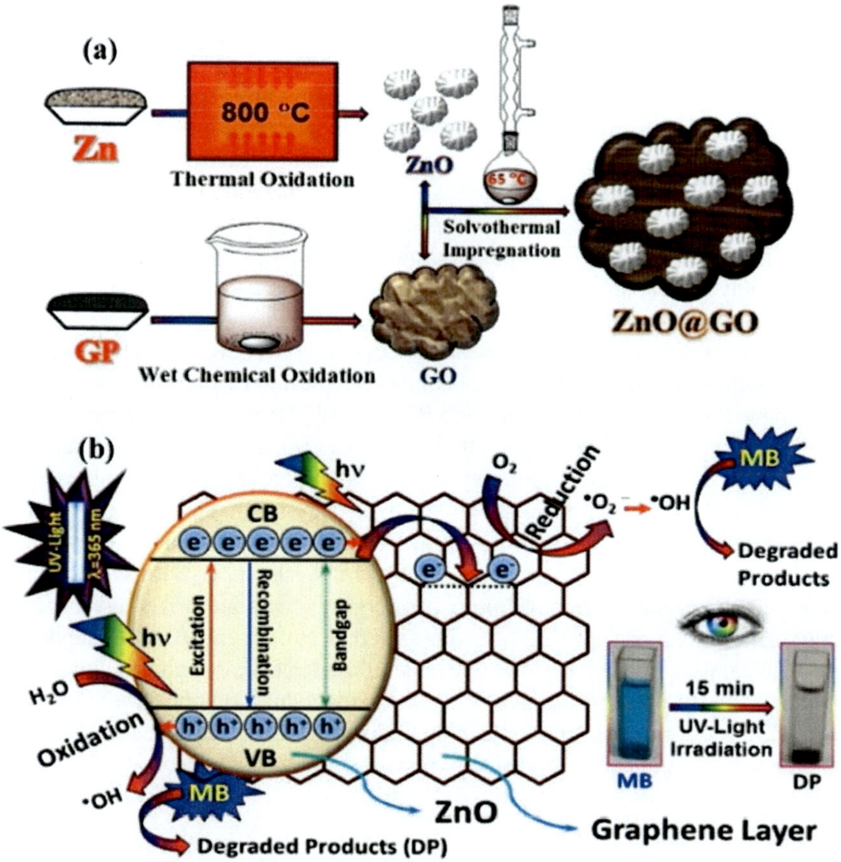

Fig. 5.12 **a** Schematically illustrated synthesis of ZnO, GO, and ZnO/GO nanocomposite, **b** photocatalysis removal reaction mechanism of methylene blue with ZnO/GO nanocomposites under ultraviolet light illumination, reproduced from Ref. [111], with permission from Elsevier, copyright (2016)

illumination of 60 min. Furthermore, the results showed that graphene oxide exhibited key role to improve photocatalysis performance of ZnO/GO for methylene blue removal as shown in schematic diagram of Fig. 5.12b.

5.2.3 *Cerium Dioxide-Based Nanomaterials*

Cerium dioxide (CeO_2) well known as ceria is potential candidate for photocatalysis decomposition of contaminants due to its nontoxic nature, higher thermal stability, and specific chemical reactivity [112, 113]. The ceria photocatalyst has a band gap of 3.1 eV, which is wide band gap. The photocatalytic efficiency CeO_2 majorly depends on heat treatment because various organophosphate pesticides have positive charge on phosphorous; thus, they can be easily adsorbed on surface of ceria surface through electrostatic attraction with the hydroxyl radicals of CeO_2 nanostructures. It was seen that the CeO_2 nanoparticles heat treated at lower calcination temperature possess a larger number of hydroxyl radicals, thereby adsorbed organophosphate pollutants could be easily degraded [114]. For instance, CeO_2-Fe_2O_3 nanocomposites exhibited fast decontamination of the organophosphate pesticides [115]. The CeO_2-Fe_2O_3 nanocomposite heat treated at 300–400 °C temperature displayed maximum decomposition performance and good magnetic characteristics for easy separation of photocatalyst to reuse. The samples calcined at high temperature than 500 °C considerably reduced the surface area, surface active sites, and pore volume of CeO_2-Fe_2O_3-based nanocomposites.

Tang et al. prepared a one-dimensional CeO_2 nanotubes (CeO_2-NT) with well-shaped hollow interior which displayed higher photocatalytic performance than that of commercial CeO_2 nanoparticles and P25 toward aromatic benzene degradation [116]. The superior photocatalytic efficiency of CeO_2 nanotubes than the nanoparticles was ascribed to the morphology-dependent properties of novel photocatalyst. Xu et al. studied the microwave-improved degradation of methyl orange dye by a heterostructure based on CuO-CeO_2 in the presence and the absence of hydrogen peroxide (H_2O_2) [113]. The produced hydroxyl radicals via H_2O_2 instantaneously increased photocatalysis removal process of organic compounds. The binary metal oxide nanostructures CeO_2-Y_2O_3 having different molar contents were applied in the disintegration of Rhodamine B dye under ultraviolet–visible photo-illumination which showed 98% removal activity higher than individual components of CeO_2 and Y_2O_3. They ascribed the superior activity to high specific surface area and enlarged oxygen vacancies in the CeO_2-Y_2O_3 composite.

5.2.4 *Iron Oxide-Based Nanomaterials*

Iron oxide (Fe_2O_3) is abundantly available n-type semiconductor with good photocatalytic ability to decompose wastewater pollutants. The easy recovery from the aqueous environment because of its magnetism (intrinsic) and low band gap energy (2.2 eV) makes it an attractive material for efficient visible-driven photocatalytic activity [117]. The iron oxide has many crystallographic forms including α-Fe_2O_3, γ-Fe_2O_3, α-FeOOH, β-FeOOH, and γ-FeOOH which were extensively utilized for the decomposition of highly toxic pollutants under the visible portion of solar spectrum [118]. Khedr et al. [119] prepared iron oxide nanoparticles of different sizes (35, 100, 150 nm) and used for the decomposition of Congo red. It was seen that the Congo red was fully decomposed when Fe_2O_2 nanoparticles of 35 nm were used as photocatalyst with and without visible-light irradiation, and no noticeable influence of light on catalyst itself decomposition was observed. In another report, Congo red dye was also degraded using α-Fe_2O_3 nanostructures synthesized via hydrothermal process. The α-Fe_2O_3 photocatalyst successfully degraded about 99% of Congo red dye in just 24 min of visible-light irradiation. Moreover, the used photocatalyst α-Fe_2O_3 was easily separated from aqueous medium for reuse [120].

Besides these advantages, iron oxide suffers from lower positive hole-diffusion length (2–4 nm) and poor electronic mobility in 0.01–0.1 cm^2/Vs^4 range, which results into fast recombination of photogenerated charge carriers within time of nanoseconds [121, 122]. These disadvantages of Fe_2O_3 nanomaterials are great challenges for their commercial use. Compared to other iron oxide phases, α-Fe_2O_3 has high stability under normal conditions [123]. Moreover, photocatalysis performance of α-Fe_2O_3 largely relies on the particle size, specific surface area, and morphology [124]. Recently, different morphologies of Fe_2O_3 such as nanoparticles, nanorods, nanowires, nanotubes, nanobelts, nanosheets, nanoflakes, and nanorings were prepared and applied for photocatalytic applications [121, 122, 124–128]. For instance, Liang et al. [128] fabricated a hybrid based on Pt/α-Fe_2O_3 nanorings as shown in Fig. 5.13. This novel α-Fe_2O_3/Pt nanorings photocatalyst exhibited superior photocatalytic efficiency in methyl orange decomposition than that of pure α-Fe_2O_3 nanorings in the same time of irradiation.

Modification of iron oxide with other semiconductor nanomaterials is another good strategy to increase photocatalysis activity via efficient charge carrier's separation and enlarges visible-light absorption. The design of composite supports the likelihood for improved performance and multi-functional characteristics as compared to single photocatalyst [129]. A low-cost direct Z-scheme g-C_3N_4/Fe2O_3-based heterojunction with increased photocatalysis removal performance for RhB was synthesized by means of simple in situ preparation strategy [130]. The direct formation of Z-scheme heterojunction not only favored the separation of charge carriers but also preserved the strong redox ability of photogenerated charges in g-C_3N_4/Fe_2O_3 which is responsible for higher photocatalysis activity. The authors proposed a photocatalysis reaction mechanism to demonstrate the enhanced degradation process of RhB when exposed to photo-irradiation using excellent g-C_3N_4/Fe2O_3 heterostructure.

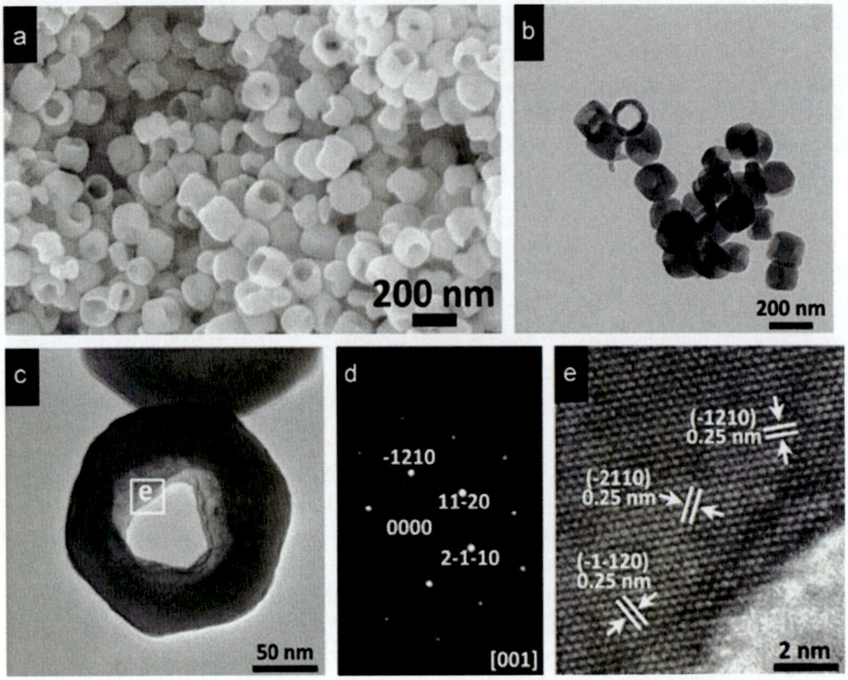

Fig. 5.13 (**a**) SEM, (**b**, **c**) TEM images, **d** SAED pattern, and **e** HRTEM image of pristine α-Fe_2O_3 nanorings, reproduced from Ref. [128] with permission from Elsevier, copyright (2014)

5.2.5 Copper Oxide-Based Nanomaterials

Copper oxide is a p-type semiconductor material with small band gap of about 1.4 eV [117, 131]. The copper oxide has two forms such as CuO and Cu_2O. Among these, the CuO has displayed higher photocatalytic efficiency than that of Cu_2O. In the recent past, CuO nanomaterials were loaded with other semiconductors including TiO_2, ZnO, SiO_2, MoS_2, and rGO, etc., which modified electronic, optical, and charge transport properties as well as photocatalytic performance of these nanomaterials [132–134]. The synthesis of three-dimensional (3D) ZnO/Fe_3O_4-CuO-based heterostructure with increased photocatalysis activity in removal of Congo red dye under visible-light illumination is studied by Malwal et al. [132]. The higher photocatalysis efficiency was ascribed to good charge separation due to the development of p–n heterostructure between n-type ZnO/Fe_3O_4 and p-type CuO semiconductors.

The p–n heterojunction material incorporated on zeolite (CuO-TiO_2-zeolite) showed superior photocatalysis efficiency in the decomposition of azo dyes under visible photo-illumination. It was observed that the composite degraded about 90% of methylene blue (MB) dye within one hour as compared to TiO_2-zeolite (84.1%,) and CuO-zeolite (85.2%) of MB dye under same reaction conditions. The reduction of electron–hole pairs recombination and enhanced interfacial charge transport due to higher active specific surface area and band gap decrease improved the photocatalysis efficiency under visible-light illumination [135]. Recently, Choi et al. [136] used highly efficient method to prepare well dispersed Cu, Cu_2O, and CuO nanoparticles on reduced graphene oxide (rGO). As compared to other photocatalysts, the CuO-rGO-based composites displayed superior photocatalytic activity and reusability toward decomposition of dye molecules when exposed to visible photoirradiation. The authors attributed superior photocatalytic performance CuO-rGO to the synergistic combination of higher dye adsorption, good electron accepting properties of rGO, the surface hydroxyl radicals in the CuO-rGO, and the smaller band gap and lower size of the CuO nanoparticles.

5.2.6 Zinc Sulfide-Based Nanomaterials

Zinc sulfide (ZnS) has direct and wide band gap of about 3.6 eV, which makes it appropriate photocatalytic nanomaterial for visible light and highly effective for UV light [137, 138]. ZnS has two crystallographic phases such as hexagonal wurtzite and cubic sphalerite [139]. Due to nontoxic nature, ZnS is an excellent photocatalyst for the removal of contaminants particularly for decomposition of organic pollutants from wastewater. ZnS has the ability to rapidly create charge carriers under photo-excitation and shows a superior physical stability under UV light irradiation. In wastewater treatment, ZnS as a photocatalyst was applied to remove organic contaminants including dyes, halogenated derivatives, and p-nitrophenol [140, 141]. However, due to wide band gap, ZnS is not an efficient photocatalyst because it can only absorb 4% (UV) of incoming solar light. Therefore, different attempts were made to enhance its photocatalytic performance via morphological modifications such as nanorods, nanotubes, nanobelts, and nanowires [142–146]. For instance, Jin et al. [142] fabricated clusters of ZnS nanorods using deposition process for ZnS and Zn on Au films at 530 °C. The prepared product contains good uniformity of zinc oxide nanorods with diameter of 200 nm diameter and length of 1–2 μm, demonstrating polar top and bottom surface.

However, overall obtained results by changing morphology were not so much encouraging because only a little shift in light spectrum was observed in high wavelength range. The modification of ZnS via doping with ions and heterojunction formation with other semiconductor displays vital contribution in improving photocatalysis efficiency of ZnS. Firstly, doping ZnS with element is a facile route for photosensitization of semiconducting photocatalyst and enhancing photocatalysis efficiency. The doping elements provide extra donor or accepter levels inside band gap of semiconducting material [147]. Therefore, it is a versatile technique to shift optical absorption response of a semiconductor photocatalyst in higher wavelength region of solar spectrum. For example, Vojtyla et al. [148] studied the iron-doped ZnS photocatalyst which showed higher photocatalysis activity in reduction of hexavalent chromium in water under visible photo-irradiation (λ > 500 nm). Secondly, ZnS-based heterostructure formation is also good strategy to lessen reunion of charge species for improved photocatalysis productivity in disintegration of water pollutants. Sanad et al. [149] prepared a ZnO-ZnS heterostructure via sol–gel method. The loading of ZnO with ZnS had stabilized the ZnO nanoparticles to reduce reunion of charge carriers. Furthermore, the photocreated holes and hydroxyl radicals participated well in increasing photocatalysis efficiency during removal of contaminants employing ZnO-ZnS nanomaterial.

5.2.7 *Cadmium Sulfide-Based Nanomaterials*

Cadmium sulfide (CdS) is a visible-light active photocatalytic material because it carries direct band gap of approximately 2.43 eV at atmospheric conditions [137, 150]. Various physical and chemical methods were used for preparation of CdS-based nanostructures. The CdS exhibits size-dependent good physical and chemical properties. Generally, CdS was used as a capable nanomaterial for wastewater applications due to its suitable band gap and other characteristics. However, higher reunion of charge species and less stability in aqueous solution are the major problems with it, which limits its application as photocatalytic in environmental remediation. Different attempts were made by scientists for getting good performance in photocatalysis, for example, doping CdS with elements and/or nanocomposite formation. Chen et al. investigated rapid disintegration of rhodamine B under visible photo-illumination employing nanospheres of CdS-carbon [151]. Malghe et al. prepared C/ZnO/CdS nanocomposites via microemulsion technique [152]. The coupling of two semiconducting materials could increase light absorptivity of photocatalyst and inhibit reunion of charges. Due to this, C/ZnO/CdS sample exhibited superior visible-light responsive photocatalytic performance than that of pure as well as C-doped ZnO and CdS as shown in Fig. 5.14a. Furthermore, the rate constant of MB decomposition using C/ZnO/CdS photocatalyst is 0.0732 min^{-1}, which is about sixfold higher than pure ZnO (0.0121 min^{-1}) as displayed in Fig. 5.14b. In addition to this, the prepared photocatalyst also demonstrated good stability of C/ZnO/CdS after three consecutive cycles (Fig. 5.14c). In another report, tantalum-doped CdS nanoparticles showed good action in disintegration of MB dye when irradiated with visible light [153]. The presence of Ta^{5+} in CdS prolonged the recombination of charge carriers via creating temporary trapping sites, which ultimately caused an improved photocatalytic efficiency. The authors proposed that dynamic species including O_2^- and hydroxyl radicals were liable in disintegration of methylene blue as shown in possible photocatalytic mechanism of Fig. 5.14d.

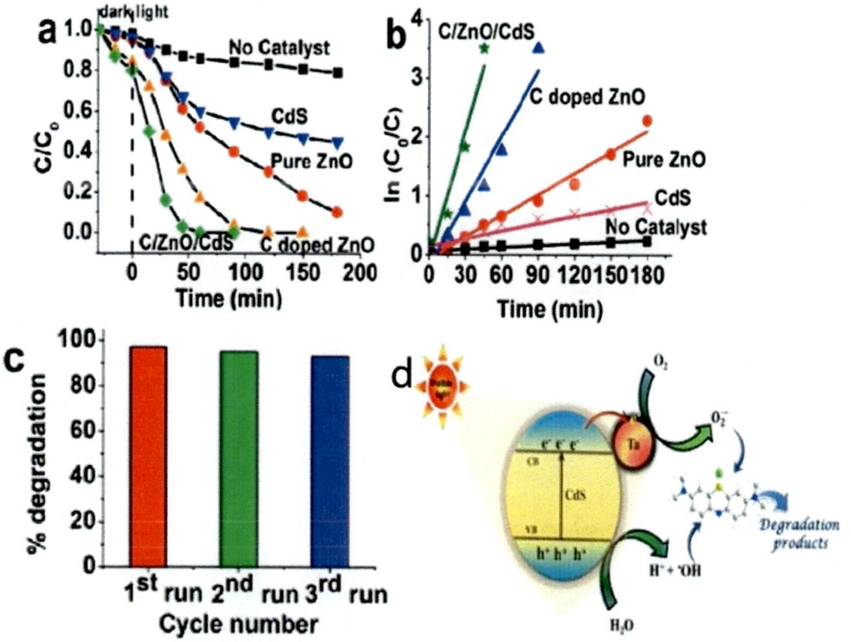

Fig. 5.14 **a** Visible-light sensitive photocatalysis removal of MB without catalyst, in existence of pristine ZnO, CdS, C-doped ZnO, and C-ZnO-CdS composite, **b** graph of ln(C0/C) vs. illumination time without photocatalyst and in the presence of bare ZnO, CdS, C-doped ZnO and C/ZnO/CdS composite, **c** cyclic runs showing photocatalysis removal of MB in the presence of C/ZnO/CdS sample [152], and **d** the possible photocatalysis mechanism of Ta-doped CdS nanoparticles, reproduced from Ref. [153] with permission from Elsevier, copyright (2019)

5.2.8 Silver Sulfide-Based Nanomaterials

Silver sulfide (Ag_2S) has direct and lower band gap (~1.1 eV) semiconducting material. It displayed good chemical stability, outstanding photo-electric and thermoelectric characteristics [154]. Ag_2S has three crystalline shapes, the natural solid bulk form including monoclinic acanthite (a-phase, stable below 179 °C), body-centered cubic (b-phase, stable above 180 °C), and a high-temperature face-centered cubic structure (g-phase, stable above 586 °C) [137, 155]. For photocatalytic applications, different Ag_2S-based nanostructures were reported. However, major problem of Ag_2S nanoparticles is their instability under atmospheric environment, because with decreasing particle size the stability of the Ag_2S nanoparticles decreases. Zhang et al. prepared a highly stable and capable Ag-Ag_2S heterodendrites (HDs) photocatalytic material using a simple electro-deposition and subsequent in situ sulfuration process [156]. The prepared photocatalyst showed outstanding performance for decomposition of methylene blue (MB) under solar irradiation. The authors attributed

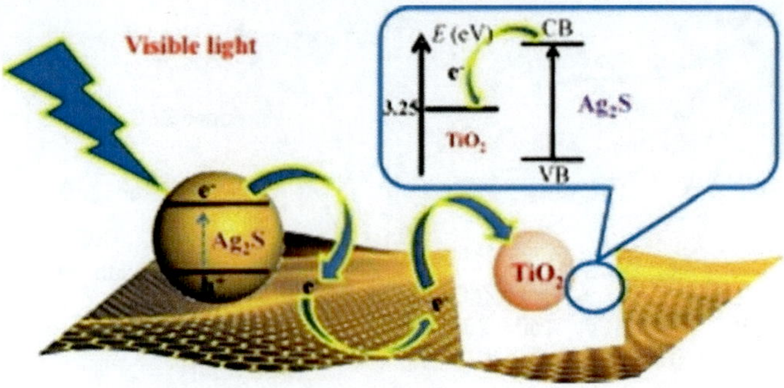

Fig. 5.15 Photo-excitation and charge transport phenomenon over TiO_2 nanoparticles, Ag_2S nanoparticles, and nanosheets of graphene, reproduced from Ref. [157] with permission from Elsevier, copyright (2013)

the improved performance to cooperative effects between Ag and Ag_2S nanomaterials including largely extended absorption of visible light, low charge transport resistance, and high separation ability for photocreated charges. Furthermore, the radical catching measurement was done to distinguish major energetic radicals in disintegration of MB dye. The authors found that main species which actively participated in photocatalytic degradation of MB employing Ag-Ag_2S hybrid were O_2^- and h^+.

Meng et al. [157] reported boosted visible-light responsive photocatalysis activity of Ag_2S-graphene-TiO_2 composites for RhB disintegration. The superior efficiency of this composite was credited to synergistic sound effects of higher charge movement and higher shift of absorption edge in visible portion. Moreover, the excitation, charge transport process between Ag_2S, TiO_2 and graphene nanosheets when exposed to visible light is shown in Fig. 5.15, where lifetimes of photocreated charges were prolonged in transportation process, thus higher photocatalytic performance by Ag_2S-graphene/TiO_2 nanocomposite.

5.2.9 *Bismuth Sulfide-Based Nanomaterials*

Bismuth sulfide (Bi_2S_3) has got a lot of attention recently due to its unique optical properties and advanced photocatalysis activities in the degradation of contaminants [158, 159]. Due to its small band gap of about 1.3 eV, many researchers have investigated the photocatalytic performance of Bi_2S_3-based nanomaterials. For instance, Chen et al. [160] adopted a facile wet chemical method to fabricate Bi_2S_3 hierarchical nanostructure and used for disintegration of methyl orange pollutant. The prepared photocatalyst showed an enhanced performance as compared to pure Bi_2S_3 due to larger specific surface area and greater hydrophilicity. In a report, Liu et al. [161] reported Bi_2S_3-ZnS/graphene nanocomposites with enhanced performance in removal of rhodamine B under visible photo-illumination. The superior activity was due to effective separation of electron–hole pairs at contact of two semiconducting materials, which facilitated transfer of charges. Wang et al. [162] prepared highly efficient and full spectrum responsive carbon quantum dots (C-dots) modified Bi_2S_3 nanotubes, where C-dots have been synthesized by facile one-step thermolysis procedure and Bi_2S_3/C-dot nanohybrids were produced via facile hydrothermal technique. A possible reaction mechanism was schematically illustrated for improved photocatalysis performance in the decomposition of methylene blue and tetracycline hydrochloride using Bi_2S_3/C-dots. As compared to pure Bi_2S_3 nanotubes, the synthesized Bi_2S_3/C-dot nanohybrid exhibited good photocatalysis efficiency for the degradation of both dyes because of well separation of electrons–holes and increased specific surface area. Vattikuti et al. [163] synthesized a nanocomposite based on Bi_2S_3 nanorods loaded by MoS_2 nanosheets using facile simple hydrothermal process without surfactants. The obtained Bi_2S_3-MoS_2 nanocomposite showed two times superior photocatalysis performance compared to pristine Bi_2S_3. The authors found that the active radicals O_2^-, h^+, and hydroxyl radicals favored the photocatalytic degradation reaction for phenol red (PR) degradation.

5.2.10 Copper Sulfide-Based Nanomaterials

Copper oxide (CuS) has extraordinary optical properties which is beneficial for photocatalytic applications. The CuS stoichiometry changes from CuS_2 to Cu_2S easily [137]. The basic crystalline structures of copper sulfide are Cu_2S and CuS. Furthermore, CuS possesses various morphologies such as microspheres, nanotubes, nanoflakes, and nanoparticles, and it has unique physical and chemical characteristics due to which it is excellent photocatalyst material [164]. Basu et al. [165] prepared hexagonal-shaped nanoplates of copper sulfide (CuS) showing good hierarchical structure like morphology (Fig. 5.16a–d). CuS in ethanol dispersion presented a band gap of 2.2 eV, which greatly assisted for visible-light absorption and enhanced photocatalysis decomposition of various contaminants under indoor light. Hosseinpour et al. [166] also prepared hierarchical structures of covellite CuS via facile chemical process. They observed an excellent photocatalysis activity of CuS hierarchical nanostructure in the decomposition of MB dye under solar photo-illumination than that of commercial CuS and TiO_2 photocatalysts.

There are few others metal sulfide photocatalysts including CoS, FeS_2, and PbS which were used for photocatalysis disintegration of contaminants. However, major issue with metal sulfide-based photocatalysts is their chemical stability in aqueous solution because they can be easily decomposed into their counterparts. In order to avoid that, advanced chemical synthetic routs are required for the preparation of efficient photocatalysts.

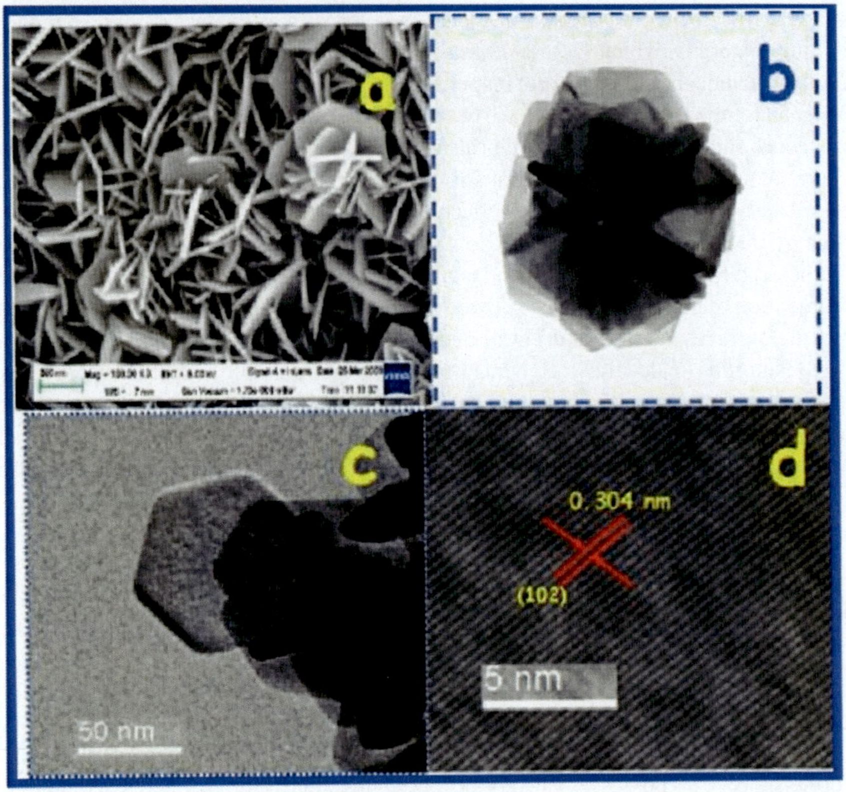

Fig. 5.16 **a** FESEM and TEM image of **b** stacked plates, **c** single hexagonal plate, d HR-TEM image of copper sulfide, reprinted from Ref. [165] with permission from ACS, copyright (2010)

References

1. Y. Boyjoo, H. Sun, J. Liu, V.K. Pareek, S. Wang, A review on photocatalysis for air treatment: from catalyst development to reactor design. Chem. Eng. J. (2016). https://doi.org/10.1016/j.cej.2016.06.090
2. H. Ren, P. Koshy, W.-F. Chen, S.Qi, C.C. Sorrell, Photocatalytic materials and technologies for air purification. J. Hazard. Mater. **325**, 340–366 (2017)
3. A. Di Paola, E. Garcia-Lopez, G. Marci, L. Palmisano, A survey of photocatalytic materials for environmental remediation. J. Hazard. Mater. **211**, 3–29 (2012)
4. J. Chen, C.-S. Poon, Photocatalytic construction and building materials: fromfundamentals to applications. Build. Environ. **44**, 1899–1906 (2009)
5. H. Ahmad, S. Kamarudin, L. Minggu, M. Kassim, Hydrogen fromphoto-catalytic water splitting process: a review. Renew. Sustain. Energy Rev. **43**, 599–610 (2015)
6. D.A. Hanaor, C.C. Sorrell, Review of the anatase to rutile phase transformation. J. Mater. Sci. **46**, 855–874 (2011)
7. Z. Liu, F.T. Chen, P.F. Fang, S.J. Wang, Y.P. Gao, F. Zheng, Y. Liu, Y.Q. Dai, Study of adsorption-assisted photocatalytic oxidation of benzene with TiO_2/SiO_2 nanocomposites. Appl. Catal. a-Gen. **451**, 120–126 (2013)

8. H.B. Huang, H.L. Huang, L. Zhang, P. Hu, X.G. Ye, D.Y.C. Leung, Enhanced degradation of gaseous benzene under vacuum ultraviolet (VUV) irradiation over TiO_2 modified by transition metals. Chem. Eng. J. **259**, 534–541 (2015)
9. J.J. Murcia, M.C. Hidalgo, J.A. Navio, V. Vaiano, D. Sannino, P. Ciambelli, Cyclohexane photocatalytic oxidation on Pt/TiO_2 catalysts. Catal. Today **209**, 164–169 (2013)
10. Y.L. Chen, X.X. Cao, B.F. Gao, B.Z. Lin, A facile approach to synthesize N-doped and oxygen-deficient TiO_2 with high visible-light activity for benzene decomposition. Mater. Lett. **94**, 154–157 (2013)
11. T.D. Pham, B.K. Lee, Novel adsorption and photocatalytic oxidation for removal of gaseous toluene by V-doped TiO_2/PU under visible light. J. Hazard. Mater. **300**, 493–503 (2015)
12. X. Qiu, M. Miyauchi, K. Sunada, M. Minoshima, M. Liu, Y. Lu, D. Li, Y. Shimodaira, Y. Hosogi, Y. Kuroda et al., Hybrid CuxO/TiO_2 nanocomposites as risk-reduction materials in indoor environments. ACS Nano **6**, 1609–1618 (2012)
13. W.-K. Jo, J.-T. Kim, Application of visible-light photocatalysis with nitrogen-doped or unmodified titanium dioxide for control of indoor-level volatile organic compounds. J. Hazard. Mater. **164**, 360–366 (2009)
14. A. Trapalis, N. Todorova, T. Giannakopoulou, N. Boukos, T. Speliotis, D. Dimotikali et al., TiO_2/graphene composite photocatalysts for NOx removal: a comparison of surfactant-stabilized graphene and reduced graphene oxide. Appl. Catal. B Environ. **180**, 637–647 (2016)
15. N. Seifvand, E. Kowsari, Novel TiO_2/graphene oxide functionalized with a cobalt complex for significant degradation of NOx and CO. RSC Adv. **5**, 93706–93716 (2015)
16. J.J. Li, S.C. Cai, E.Q. Yu, B. Weng, X. Chen, J. Chen, H.P. Jia, Y.J. Xu, E_cient infrared light promoted degradation of volatile organic compounds over photo-thermal responsive Pt-rGO-TiO2composites. Appl. Catal. B Environ. **233**, 260–271 (2018)
17. A. Pastor, J. Balbuena, M. Cruz-Yusta, I. Pavlovic, L. Sánchez, ZnO on rice husk: A sustainable photocatalyst for urban air purification. Chem. Eng. J. **368**, 659–667 (2019)
18. E. Kowsari, B. Bazri, Synthesis of rose-like ZnO hierarchical nanostructures in the presenceof ionic liquid/Mg_2+ for air purification and their shape-dependentphotodegradation of SO_2, NOx, and CO. Appl. Catal. Gen. **475**, 325–334 (2014)
19. P. Nagaraju, S.H. Puttaiah, K. Wantala, B. Shahmoradi, Preparation of modified ZnO nanoparticles for photocatalytic degradation of chlorobenzene, Appl. Water Sci. **10**, 137 (2020)
20. E. Kowsari, S. Abdpour, Investigation performance of rodrod ZnO/CdO composites, synthesized in ionic liquid medium as photocatalytic for degradation of air pollutants (SO_2 and NO_X), Optik – Int. J. Light Electron Opt. http://dx.doi.org/10.1016/j.ijleo.2016.09.084
21. T. Fukumura, E. Sambandan, H. Yamashita, Synthesis and VOC degradation ability of a CeO_2/WO_3 thin-layer visible-light photocatalyst. Mater. Res. Bull. **94**, 493–499 (2017)
22. Y. Wicaksana, S. L, Jason Scott, R. Amal, Tungsten trioxide as a visible light photocatalyst for volatile organic carbon removal. Molecules **19**, 17747–17762 (2014)
23. I. Jansson, K. Yoshiiri, H. Hori, F.J. García-García, S. Rojas, B. Sánchez, B. Ohtani, S. Suárez, Visible light responsive Zeolite/WO3–Pt hybrid photocatalysts fordegradation of pollutants in air, Appl. Catal. Gen. (2016)
24. M. Mishra, D.-M. Chun, a-Fe_0O3 as a photocatalytic material: a review. Appl. Catal. A Gen. **498**, 126–141 (2015)
25. J. Balbuena, M. Cruz-Yusta, A. Cuevas, F. Martín, A. Pastor, R. Romero, L. Sanchez, Hematite porous architectures as enhanced air purification photocatalyst. J. Alloy. Compd. **797**, 166–173 (2019)
26. Z.-F. Huang, L. Pan, J.-J. Zou, X. Zhang, L. Wang, Nanostructured bismuthvanadate-based materials for solar-energy-driven water oxidation: a reviewon recent progress. Nanoscale (2014)
27. H. Fan, T. Jiang, H. Li, D. Wang, L. Wang, J. Zhai, D. He, P. Wang, T. Xie, Effectof $BiVO_4$ crystalline phases on the photoinduced carriers behavior and photocatalytic activity. J. Phys. Chem. C **116**, 2425–2430 (2012)

28. J. Yang, Q. Shi, R. Zhang, M. Xie, X. Jiang, F. Wang, X. Cheng, W. Han, $BiVO_4$ quantum tubes loaded on reduced graphene oxide aerogel as efficient photocatalyst for gaseous formaldehyde degradation. Carbon (2018). https://doi.org/10.1016/j.carbon.2018.06.003
29. T. Alammar, I. Hamm, M. Wark, A.-V. Mudring, Low-temperature route tometal titanate perovskite nanoparticles for photocatalytic applications. Appl. Catal. B (2014)
30. S. Thirumalairajan, K. Girija, N.Y. Hebalkar, D. Mangalaraj, C. Viswanathan, N. Ponpandian, Shape evolution of perovskite $LaFeO_3$ nanostructures: asystematic investigation of growth mechanism, properties and morphologydependent photocatalytic activities. RSC Adv. **3**, 7549–7561 (2013)
31. Y. Li, S. Yao, W. Wen, L. Xue, Y. Yan, Sol–gel combustion synthesis andvisible-light-driven photocatalytic property of perovskite $LaNiO_3$. J. Alloys Compd. **491**, 560–564 (2010)
32. H. Li, S. Yin, Y. Wang, T. Sekino, S.W. Lee, T. Sato, Roles of Cr_3+ doping andoxygen vacancies in $SrTiO_3$ photocatalysts with high visible light activity for NO removal. J. Catal. **297**, 65–69 (2013)
33. U. Sulaeman, S. Yin, T. Sato, Solvothermal synthesis and photocatalyticproperties of chromium-doped $SrTiO_3$ nanoparticles. Appl. Catal. B **105**, 206–210 (2011)
34. U.G. Akpan, B.H. Hameed, Parameters affecting the photocatalytic degradation of dyes using TiO_2-based photocatalysts: a review. J. Hazard. Mater. **170**, 520–529 (2009)
35. J. Schneider, M. Matsuoka, M. Takeuchi, J. Zhang, Y. Horiuchi, M. Anpo et al., Understanding TiO_2 photocatalysis: mechanisms and materials. Chem. Rev. **114**, 9919–9986 (2014)
36. T. Aarthi, P. Narahari, G. Madras, Photocatalytic degradation of Azure and Sudan dyes using nano TiO_2. J. Hazard. Mater. **149**, 725–734 (2007)
37. H.S. Kim, J.W. Lee, N. Yantara, P.P. Boix, S.A. Kulkarni, S. Mhaisalkar et al., High efficiency solid-state sensitized solar cell-based on submicrometer rutile TiO_2 nanorod and CH3NH3PbI3 perovskite sensitizer. Nano Lett. **13**, 2412–2417 (2013)
38. Pan, L., Zou, J.-J., Zhang, X., Wang, L., Photoisomerization of norbornadiene to quadricyclane using transition metal doped TiO_2. J. Ind. Eng. Chem. Res. **49**(18), 8526e8531 (2010)
39. J. Liu, X. Yu, Q. Liu, R. Liu, X. Shang, S. Zhang, W. Li, W. Zheng, G. Zhang, H. Cao, Surface-phase junctions of branched TiO_2 nanorod arrays for efficient photoelectrochemical water splitting. J Appl. Catal. Environ. **158**, 296–300 (2014)
40. S. Zhang, The microscopic origin of the doping limits in semiconductors and wide-gap materials and recent developments in overcoming these limits: a review. J. Phys. Condens. Matter **14**(34), R881 (2002)
41. K.E. Karakitsou, X.E. Verykios, Effects of altervalent cation doping of TiO_2 on its performance as a photocatalyst for water cleavage. J. Phys. Chem. **97**, 1184–1189 (1993)
42. W. Mu, J.M. Herrmann, P. Pichat, Room temperature photocatalytic oxidation of liquid cyclohexane into cyclohexanone over neat and modified TiO_2. Catal. Lett. **3**, 73–84 (1989)
43. L. Sun, J. Li, C.L. Wang, S.F. Li, H.B. Chen, C.J. Lin, An electrochemical strategy of doping Fe_3+ into TiO_2 nanotube array films for enhancement in photocatalytic activity. Sol. Energy Mater. Sol. Cells **93**, 1875–1880 (2009)
44. C.M. Teh, A.R. Mohamed, Role of titanium dioxide and iondoped titanium dioxide on phtotocatalytic degradation of organic pollutants (phenol compounds and dyes) in aqueous solutions: A review. J. Alloys Compd. **509**, 1648–1660 (2011)
45. X. Zhang, Y. Cao, S. Kan, Y. Chen, J. Tang, H. Jin, Y. Bai, L. Xiao, T. Li, B. Li, Study on the photo-induced interfacial charge transfer in TiO_2/Fe_2O_3 heterostructured composite film. Thin Solid Films **327–329**, 568–570 (1998)
46. M.C. Wang, H.J. Lin, T.S. Yang, Characteristics and optical properties of iron ion (Fe_3+)-doped titanium oxide thin films prepared by sol-gel spin coating. J. Alloys Compd. **473**, 394–400 (2009)
47. Swati Sood, Ahmad Umar, S.K. Mehta, S.K. Kansal, Highly effective Fe-doped TiO_2 nanoparticles photocatalysts for visiblelight driven photocatalytic degradation of toxic organic compounds. J. Colloid Interface Sci. **450**, 213–223 (2015)
48. R. Daghrir, P. Drogui, D. Robert, Modified TiO_2 for environmental photocatalytic applications: a review. Ind. Eng. Chem. Res. **52**, 3581–3599 (2013)

49. H. Wang, J. P. Lewis, Effects of dopant states on photoactivity in carbon-doped TiO_2. J. Phys. Condens. Matter. **17**, L209−L213 (2005)
50. Y. Choi, T. Umebayashi, M. Yoshikawa, Fabrication and characterization of C-doped anatase TiO_2 photocatalyst. J. Mater. Sci. **39**, 1837–1839 (2004)
51. C.D. Valentin, E. Finazzi, G. Pacchioni, A. Selloni, S. Livraghi, M.C. Paganini, E. Giamello, N-doped TiO_2: theory and experiment. Chem. Phys. **339**, 44–56 (2007)
52. R. Asahi, T. Morikawa, T. Ohwaki, K. Aoki, Y. Taga, Visible light photocatalysis in nitrogen-doped titanium oxides. Science **293**(5528), 269–271 (2001)
53. S. Yang, L. Gao, New method to prepare nitrogen-doped titanium dioxide and its photocatalytic activities irradiated by visible light. J. Am. Ceram. Soc. **87**, 1803–1805 (2008)
54. T. Ohno, T. Mitsui, M. Matsumura, Photocatalytic activity of S-doped TiO2 photocatalyst under visible light. Chem. Lett. **32**, 364–365 (2003)
55. T. Umebayashi, T. Yamaki, S. Tanaka, K. Asai, Visible light induced degradation of methylene blue on S-doped TiO_2. Chem. Lett. **32**, 330–331 (2003)
56. X. Hong, Z. Wang, W. Cai, F. Lu, J. Zhang, Y. Yang, N. Ma, Y. Liu, Visible light activated nanoparticle photocatalyst of iodine doped titanium dioxide. Chem. Mater. **17**, 1548–1552 (2005)
57. Y. N. Tan, C. L. Wong, A. R. Mohamed, An overview on the photocatalytic activity of nano-doped TiO_2 in the degradation of organic pollutants. ISRN Mater. Sci. ID261219, 1−18 (2011)
58. W. Ren, Z. Ai, F. Jia, L. Zhang, X. Fan, Z. Zou, Low temperature preparation and visible light photocatalytic activity of mesoporous carbon-doped crystalline TiO_2. Appl. Catal. B **69**, 138−144 (2007)
59. J.C. Yu, J. Yu, W. Ho, Z. Jiang, L. Zhang, Effects of Fdoping on the photocatalytic activity and microstructures of nanocristalline TiO_2 powders. Chem. Mater. **14**, 3808–3816 (2002)
60. Y. Liu, J. Liu, Y. Lin, Y. Zhang, Y. Wei, Simple fabrication and photocatalytic activity of S-doped TiO_2 under low power LED visible light irradiation. Ceram. Int. **35**, 3061–3065 (2009)
61. D. Chen, Z. Jiang, J. Geng, Q. Wang, D. Yang, Carbon and nitrogen Co-doped TiO_2 with enhanced visible light photocatalytic activity. Ind. Eng. Chem. Res. **46**, 2741–2746 (2007)
62. J. Saien, Z. Mesgari, Highly efficient visible-light photocatalyst of nitrogen-doped TiO2 nanoparticles sensitized by hematoporphyrin. J. Mol. Catal. A Chem. http://dx.doi.org/10.1016/j.molcata.2015.12.027
63. L. Zhang, M. S. Tse, O. K. Tan, Y. X. Wang, M. Han, Facile fabrication and characterization of multi-type carbon-doped TiO_2 for visible light-activated photocatalytic mineralization of gaseous toluene. J. Mater. Chem. A **1**, 4497 (2013)
64. H. Tada, T. Kiyonaga, S.-I. Naya, Rational design and applications of highly efficient reaction systems photocatalyzed by noble metal nanoparticle-loaded Titanium(IV) dioxide. Chem. Soc. Rev. **38**, 1849–1858 (2009)
65. P.V., Photoinduced transformations in semiconductor metal nanocomposite assemblies. Pure Appl. Chem. **74**, 1693–1706 (2002a)
66. M. El-Kemary, Y. Abdel-Moneam, M. Madkour, I. El-Mehasseb, Enhanced photocatalytic degradation of safranin-O by heterogeneous nanoparticles for environmental applications. J. Lumines. **131**, 570–576 (2011)
67. A. Ayati, A. Ahmadpour, F. Bamoharram, B. Tanhaei, M. Manttari, M. Sillanpaa, A review on catalytic applications of Au/TiO_2 nanoparticles in the removal of water pollutant. Chemosphere (2014)
68. Y. Chen, J. Bian, L. Qi, E. Liu, J. Fan, Efficient degradation of methylene blue over two-dimensional Au/TiO_2 nanosheet films with overlapped light harvesting nanostructures. J. Nanomaterials (2015). https://doi.org/10.1155/2015/905259
69. D. Zhua, Q. Zhou, Action and mechanism of semiconductor photocatalysis on degradation of organic pollutants in water treatment: a review. Environ. Nanotechnol. Monit. Manage. **12**, 100255 (2019)

70. S. Ullah, R. A.M. Butt, N.D. Gohar, Strategies of making TiO_2 and ZnO visible light active. J. Hazard. Mater. **170**(2–3), 560–569 (2009)
71. X. Zheng, D. Li, X. Li, J. Chen, C. Cao, J. Fang, J. Wang, Y. He, Y. Zheng, Construction of ZnO/TiO_2 photonic crystal heterostructures for enhanced photocatalytic properties. Appl. Catal. B: Environ. **168**, 408–415 (2015)
72. S. Bai, H. Liu, J. Sun, Y. Tian, S. Chen, J. Song, R. Luo, D. Li, A. Chen, C.C. Liu, Improvement of TiO_2 photocatalytic properties under visible light by WO_3/TiO_2 and MoO_3/TiO_2 composites. Appl. Surf. Sci. **338**, 61–68 (2015)
73. V. Dutta, P. Singh, P. Shandilya et al., Review on advances in photocatalytic water disinfection utilizing graphene and graphene derivatives-based nanocomposites. J. Environ. Chem. Eng. **7**, 103132 (2019)
74. N.R. Khalid, E. Ahmed, Y. Zhang, M. Ahmad, Z. Hong, Enhanced photocatalytic activity of graphene-TiO_2 composite under visible light irradiation. Curr. Appl. Phys. **13**, 659–663 (IF-2.026)
75. M. Faraldos, A. Bahamonde, Environmental applications of titania-graphene photocatalysts. Catal. Today **285**, 13–28 (2017)
76. C. Chen, W. Cai, M. Long, B. Zhou, Y. Wu, D. Wu, Y. Feng, ACS Nano **11**, 6425–6432 (2010)
77. H. Zhang, X. Lv, Y. Li, Y. Wang, J. Li, ACS Nano **4**(1), 380–386 (2010)
78. N.R. Khalid, Z. Hong, E. Ahmed, Y. Zhang, H. Chan, M. Ahmad, Synergistic effects of Fe and graphene on photocatalytic activity enhancement of TiO_2 under visible light. Appl. Surf. Sci. **258**, 5827–5834 (2012)
79. Y. Tang, G. Zhang, C. Liu, S. Luo, X. Xu, Magnetic TiO_2–graphene composite as high-performance and recyclable platform for efficient photocatalytic removal of herbicides from water. J. Hazard. Mater. **253**, 115–122 (2013)
80. N.R. Khalid, E. Ahmed, Z. Hong, Y. Zhang, M. Ahmad, Nitrogen doped TiO_2 nanoparticles decorated on graphene sheets for photocatalysis applications. Curr. Appl. Phys. **12**, 1485–1492 (IF-1.814)
81. P. Gao, Z. Liu, M. Tai, D.D. Sun, W. Ng, Multifunctional graphene oxide-TiO_2 microsphere hierarchical membrane for clean water production. Appl. Catal. B Environ. 138–139 (2013)
82. Z. Hong, Y. Zhang, X. Jiao, C. He, N.R. Khalid, J. Yang, Solvothermal preparation of carbon-enhanced TiO_2/graphene composite and its visible light photocatalytic properties. Integr. Feroelectric, **138**, 1–7
83. N.R. Khalid, E. Ahmed, Z. Hong, M. Ahmad, Synthesis and photocatalytic properties of visible light responsive La/TiO_2-graphene composites, Appl. Surf. Sci. **263**, 254–259
84. K. Li, J. Xiong, T. Chen, L. Yan, Y. Dai, D. Song, Y. Lv, Z. Zeng, Preparation of graphene/TiO_2 composites by nonionic surfactant strategy and their simulated sunlight and visible light photocatalytic activity towards representative aqueous POPs degradation. J. Hazard. Mater. **250**, 19–28 (2013)
85. N.R. Khalid, E. Ahmed, Z. Hong, Y. Zhang, M. Ullah, M. Ahmad, Graphene modified Nd/TiO_2 photocatalyst for methyl orange degradation under visible light irradiation, Ceramic Int. **39**, 3569–3575
86. Q. Huang, S. Tian, D. Zeng, X. Wang, W. Song, Y. Li, W. Xiao, C. Xie, ACS Catal. **3**, 1477–1485 (2013)
87. H. Zhang, L.-H. Guo, D. Wang, L. Zhao, B. Wan, A.C.S. Appl, Mater. Interfaces **7**, 1816–1823 (2015)
88. M. Aleksandrzak, P. Adamski, W. Kukułka, B. Zielinska, E. Mijowska, Appl. Surf. Sci. **331**, 193–199 (2015)
89. N. Sakai, K. Kamanaka, T. Sasaki, J. Phys. Chem. C (2016). https://doi.org/10.1021/acs.jpcc.6b06842
90. Z.Y. Zhang, F. Xiao, Y.L. Guo, S. Wang, Y.Q. Liu, One-pot self-assembled three-dimensional TiO_2-graphene hydrogel with improved adsorption capacities and photocatalytic and electrochemical activities. ACS Appl. Mater. Interfaces. **5**, 2227–2233 (2013)
91. R. Giovannetti, E. Rommozzi, M. Zannotti, Chiara anna D'Amato, recent advances in graphene based TiO_2 Nanocomposites ($GTiO_2Ns$) for photocatalytic degradation of synthetic dyes. Catalysts **7**, 305 (2017). https://doi.org/10.3390/catal7100305

92. K.M. Lee, C.W. Lai, K.S. Ngai, J.C. Juan, Recent developments of zinc oxide based photocatalyst in water treatment technology: a review. Water Res. **88**, 428–448 (2016)
93. S. Maiti, S. Pal, K.K. Chattopadhyay, Recent advances in low temperature, solution processed morphology tailored ZnO nanoarchitectures for electron emission and photocatalysis applications. Cryst. Eng. Comm. **17**, 9264–9295 (2015)
94. S. Sakthivel, B. Neppolian, M.V. Shankar, B. Arabindoo, M. Palanichamy, V. Murugesan, Solar photocatalytic degradation of azo dye: comparison of photocatalytic efficiency of ZnO and TiO_2. Sol. Energy Mat. Sol Cells **77**, 65–82 (2003)
95. X. Chen, F. Zhang, Q. Wang, X. Han, X. Li, J. Liu et al., The synthesis of ZnO/SnO_2 porous nanofibers for dye adsorption and degradation. Dalton Trans. **44**, 3034–3042 (2015)
96. O. Yayapao, T. Thongtem, A. Phuruangrat, S. Thongtem, Ultrasonic-assisted synthesis of Nd-doped ZnO for photocatalysis. Mater. Lett. **90**, 83–86 (2013)
97. M. Khatamian, A.A. Khandar, B. Divband, M. Haghighi, S. Ebrahimiasl, Heterogeneous photocatalytic degradation of 4-nitrophenol in aqueous suspension by Ln (La^{3+} , Nd^{3+} or Sm^{3+}) doped ZnO nanoparticles. J. Mol. Catal. Chem. **365**, 120–127 (2012)
98. Y. Zheng, C. Chen, Y. Zhan, X. Lin, Q. Zheng, K. Wei, J. Zhu, Photocatalytic activity of Ag/ZnO heterostructure nanocatalyst: correlation between structure and property. J. Phys. Chem. C **112**, 10773–10777 (2008)
99. J. Liqiang, W. Baiqi, X. Baifu, L. Shudan, S. Keying, C. Weimin, F. Honggang, Investigations on the surface modification of ZnO nanoparticle photocatalyst by depositing Pd. J. Solid State Chem. **177**, 4221–4227 (2004)
100. H. Zeng, P. Liu, W. Cai, S. Yang, X. Xu, Controllable Pt/ZnO porous nanocages with improved photocatalytic activity. J. Phys. Chem. C **112**, 19620–19624 (2008)
101. P. Li, Z. Wei, T. Wu, Q. Peng, Y. Li, Au-ZnO hybrid nanopyramids and their photocatalytic properties. J. Am. Chem. Soc. **133**, 5660–5663 (2011)
102. J. Mani, H. Sakeek, S. Habouti, M. Dietze, M. Es-Souni, Macro-meso-porous TiO_2, ZnO and, ZnO–TiO_2-composite thick films. Properties and application to photocatalysis, Catal. Sci. Technol. **2**, 379–385 (2012)
103. M.T. Uddin, Y. Nicolas, C. Olivier, T. Toupance, L. Servant, M.M. Müller, H.J. Kleebe, J. Ziegler, W. Jaegermann, Nanostructured SnO_2–ZnO heterojunction photocatalysts showing enhanced photocatalytic activity for the degradation of organic dyes. Inorg. Chem. **51**, 7764–7773 (2012)
104. V. Vaiano, M. Matarangolo, O. Sacco, D. Sannino, Appl. Catal. B **209**, 621–630 (2017)
105. U. Alam, A. Khan, D. Ali, D. Bahnemann, M. Muneer, Comparative photocatalytic activity of sol–gel derived rare earth metal (La, Nd, Sm and Dy)-doped ZnO photocatalysts for degradation of dyes. RSC Adv. **8**, 17582–17594 (2018)
106. T. Lv, L. Pan, X. Liu, T. Lu, G. Zhu, Z. Sun, Enhanced photocatalytic degradation of methylene blue by ZnO-reduced graphene oxide composite synthesized via microwave-assisted reaction. J. Alloys Compd. **509**, 10086–10091 (2011)
107. P. Raizada, A. Sudhaik, P. Singh, Photocatalytic water decontamination using graphene and ZnO coupled photocatalysts: a review. Mater. Sci. Energy Technol. **2**, 509–525 (2019)
108. B. Li, T. Liu, Y. Wang, Z. Wang, ZnO/graphene-oxide nanocomposite with remarkably enhanced visible-light-driven photocatalytic performance. J. Colloid Interface Sci. **377**, 114–121 (2012)
109. R. Atchudan, T.N.J.I. Edison, S. Perumal, M. Shanmugam, Y.R. Lee, Direct solvothermal synthesis of zinc oxide nanoparticle decorated graphene oxide nanocomposite for efficient photodegradation of azo-dyes. J. Photochem. Photobio. A **337**, 100–111 (2017)
110. V.R. Posa, V. Annavaram, J.R. Koduru, V.R. Ammireddy, A.R. Somala, Graphene-ZnO nanocomposite for highly efficient photocatalytic degradation of methyl orange dye under solar light irradiation. Korean J. Chem. Eng. **33**, 456–464 (2016)
111. R. Atchudan, T.N.J.I. Edison, S. Perumal, D. Karthikeyan, Y.R. Lee, Facile synthesis of zinc oxide nanoparticles decorated graphene oxide composite via simple solvothermal route and their photocatalytic activity on methylene blue degradation. J. Photochem. Photobiol. B **162**, 500–510 (2016)

112. D.G. Shchukin, R.A. Caruso, Template synthesis and photocatalytic properties of porous metal oxide spheres formed by nanoparticle infiltration. Chem. Mater. **16**, 2287–2292 (2004)
113. D. Xu, F. Cheng, Q. Lu, P. Dai, Microwave enhanced catalytic degradation ofmethyl orange in aqueous solution over CuO/CeO_2 catalyst in the absence and presence of H2O2. Ind. Eng. Chem. Res. **53**, 2625–2632 (2014)
114. P. Janos, P. Kuran, M. Kormunda, V. Stengl, T.M. Grygar, M. Dosek et al., Cerium-dioxide as a new reactive sorbent for fast degradation of parathion methyl and some other organophosphates. J. Rare Earths **32**, 360–370 (2014)
115. P. Janoš, P. Kuráň, V. Pilařová, J. Trögl, M. Šťastný, O. Pelant et al., Magnetically separable reactive sorbent based on the $CeO_2/\gamma\text{-}Fe_2O_3$ composite and its utilization for rapid degradation of the organophosphate pesticide parathion methyl and certain nerve agents. Chem. Eng. J. **262**, 747–755 (2015)
116. Z.-R. Tang, Y. Zhang, Y.-J. Xu, A facile and high-yield approach to synthesize onedimensional CeO_2 nanotubes with well-shaped hollow interior as a photocatalyst for degradation of toxic pollutants. RSC Adv. **1**, 1772 (2011)
117. R. Gusain, K. Gupta, P. Joshi, O. Khatri, Adsorptive removal and photocatalytic degradation of organic pollutants using metal oxides and their composites: a comprehensive review. Adv. Coll. Interface. Sci. **272**, 102009 (2019)
118. O. Akhavan, R. Azimirad, Photocatalytic property of Fe_2O_3 nanograin chains coated by TiO_2 nanolayer in visible light irradiation. Appl. Catal. A Gen. **369**, 77–82 (2009)
119. W. Du, Y. Xu, Y. Wang, Photoinduced degradation of orange II on different iron (hydr) oxides in aqueous suspension: rate enhancement on addition of hydrogen peroxide, silver nitrate, and sodium fluoride. Langmuir **24**, 175–181 (2008)
120. Y. Dong, L. Xing, F. Hu, A. Umar, X. Wu, Efficient removal of organic dyesmolecules by grain-like $\alpha\text{-}Fe_2O_3$ nanostructures under visible light irradiation. Vacuum **150**, 35–40 (2018)
121. Z. Zhang, M.F. Hossain, T. Takahashi, Self-assembled hematite ($\alpha\text{-}Fe_2O_3$) nanotube arrays for photoelectrocatalytic degradation of azo dye under simulated solar light irradiation. Appl. Catal. B **95**, 423–429 (2010)
122. P. Xu, G.M. Zeng, D.L. Huang, C.L. Feng, S. Hu, M.H. Zhao, C. Lai, Z. Wei, C. Huang, G.X. Xie, Z.F. Liu, Use of iron oxide nanomaterials in wastewater treatment: a review. Sci. Total Environ. **424**, 1–10 (2012)
123. L. Guo, F. Chen, X. Fan, W. Cai, J. Zhang, S-doped-Fe_2O_3 as a highly active heterogeneous Fenton-like catalyst towards the degradation of acid orange 7 and phenol. Appl. Catal. B **96**, 162–168 (2010)
124. H. Wu, L. Wang, Phase transformation-induced crystal plane effect of iron oxide micropine dendrites on gaseous toluene photocatalytic oxidation. Appl. Surf. Sci. **288**, 398–404 (2014)
125. D. Wang, P. Yang, Y. Zhu, Growth of Fe_3O_4 nanoparticles with tunable sizes and morphologies using organic amine. Mater. Res. Bull. **49**, 514–520 (2014)
126. S.R. Kumar, M.M. Raja, D. Mangalaraj, C. Viswanathan, N. Ponpandian, Surfactant free solvothermal synthesis of monodispersed 3D hierarchical Fe_3O_4 microspheres. Mater. Lett. **110**, 98–101 (2013)
127. J.S. Xu, Y.J. Zhu, !-Fe_2O_3 hierarchically hollow microspheres self-assembled with nanosheets: Surfactant-free solvothermal synthesis, magnetic and photocatalytic properties. Cryst. Eng. Comm. **13**, 5162–5169 (2011)
128. H. Liang, X. Jiang, W. Chen, S. Wang, B. Xu, Z. Wang, $\alpha\text{-}Fe_2O_3$/Pt hybrid nanorings and their enhanced photocatalytic activities. Ceram. Int. **40**, 5653–5658 (2014)
129. H. Kim, M. Achermann, L.P. Balet, J.A. Hollingsworth, V.I. Klimov, Synthesis and characterization of Co/CdSe core/shell nanocomposites: Bifunctional magnetic-optical nanocrystals, J. Am. Chem. Soc. **127**, 544–546 (2005)
130. J. Wang, X. Zuo, W. Cai, J. Sun, X. Ge, H. Zhao, Dalton Trans. (2018). https://doi.org/10.1039/C8DT02893A
131. A.A. Umar, M. Oyama, A seed-mediated growth method for vertical array of singlecrystalline CuO nanowires on surfaces. Cryst. Growth Des. **7**, 2404–2409 (2007)

132. D. Malwal, P. Gopinath, Enhanced photocatalytic activity of hierarchical three dimensional metal oxide@CuO nanostructures towards the degradation of Congo red dye under solar radiation. Cat. Sci. Technol. **6**, 4458–4472 (2016)
133. S. Jung, K. Yong, Fabrication of CuO-ZnO nanowires on a stainless steel mesh for highly efficient photocatalytic applications. Chem. Commun. (Camb.) **47**, 2643–2645 (2011)
134. R. Gusain, P. Kumar, O.P. Sharma, S.L. Jain, O.P. Khatri, Reduced graphene oxide–CuO nanocomposites for photocatalytic conversion of CO_2 into methanol under visible light irradiation. Appl Catal B Environ **181**, 352–362 (2016)
135. L. Zhao, T. Cui, Y. Li, B. Wang, J. Han, L. Han et al., Efficient visible light photocatalytic activity of p–n junction CuO/TiO_2 loaded on natural zeolite. RSC Adv. **5**, 64495–64502 (2015)
136. J. Choi, H. Oh, S.-W. Han, S. Ahn, J. Noh, J.B. Park, Preparation and characterization of graphene oxide supported Cu, Cu_2O, and CuO nanocomposites and their high photocatalytic activity for organic dye molecule. Curr. Appl. Phys. (2016). https://doi.org/10.1016/j.cap.2016.11.020
137. D.A.G. Veerabhadram, A review on recent advances in photodegradation of dyes using doped and heterojunction based semiconductor metal sulfide nanostructures for environmental protection. Materials Today Energy **9**, 83–113 (2018)
138. S. Sur, Z. Ozturk, M. Oztas, M. Bedir, Phys. Scr. **82**, 045604 (2010)
139. S.V. Sharma, P. Chand, Acta Phys. Pol., A **81**, 653–658 (1992)
140. J. Lee, S. Ham, D.J. Jang, J. Environ. Chem. Eng. **5**, 4431–4440 (2017)
141. H. Fujiwara, H. Hosokawa, K. Murakoshi, Y. Wada, S. Yanagida, Langmuir **14**, 5154–5159 (1998)
142. C. Jin, Y. Cheng, X. Zhang, W. Zhong, Y. Deng, AC. Au, X.W.Y. Du, Catalytic growth of clusters of wurtzite ZnS nanorods through co-deposition of ZnS and Zn on Au film, CrystEngComm, **11**, 2260–2263 (2009)
143. Z. Chen, R. Wu, H. Wang, Y. Jiang, L. Jin, Y. Guo, Y. Song, F. Fang, D. Sun, Chem. Eng. J. **326**, 680–690 (2017)
144. L.W. Yin, Y. Bando, J.H. Zhan, M.S. Li, D. Golberg, Adv. Mater. **17**, 1972–1977 (2005)
145. S. Ham, D.J. Jang, J. Environ. Chem. Eng. **6**, 228–235 (2018)
146. J. Liu, Z. Guo, Y. Jia, F. Meng, T. Luo, J. Liu, J. Cryst, Growth **311**, 1423–1429 (2009)
147. G.-J. Lee, J.J. Wu, Recent developments in ZnS photocatalysts from synthesis to photocatalytic applications d a review. Powder Technol. **318**, 8–22 (2017)
148. S. Wojtyła, T. Baran, Insight on doped ZnS and its activity towards photocatalytic removing of Cr(VI) from wastewater in the presence of organic pollutants. Mater. Chem. Phys. **212**, 103–112 (2018)
149. M.F. Sanad, A.E. Shalan, S.M. Bazid, S.M. Abdelbasir, Pollutant degradation of different organic dyes using the photocatalytic activity of ZnO@ZnS nanocomposite materials. J. Environ. Chem. Eng. https://doi.org/10.1016/j.jece.2018.05.035
150. R.G. Solanki, P. Rajaram, Nano-Struct. Nano-Objects **12**, 157–165 (2017)
151. J. Chen, F. Zhang, Y.L. Zhao, Y.C. Guo, P. Gong, Z.Q. Li, H.S. Qian, Appl. Surf. Sci. **362**, 126–131 (2016)
152. Y.S. Malghe, A.B. Lavand, Synthesis Of C/ZnO/CdS Nanocomposite With Enhanced Visible Light Photocatalytic Activity. Adv. Mater. Lett. **7**(3), 239–245 (2016)
153. R. Rajendrana, K. Varadharajana, V. Jayaraman, Fabrication of tantalum doped CdS nanoparticles for enhanced photocatalytic degradation of organic dye under visible light exposure. Colloids Surf. A **580**, 123688 (2019)
154. S.I. Sadovnikov, Y.V. Kuznetsova, A.A. Rempel, Nano-Struct. Nano-Objects **7**, 81–91 (2016)
155. Y. Zhang, J. Xia, C. Li, G. Zhou, W. Yang, D. Wang, H. Zheng, Y. Du, X. Li, Q. Li, J. Mater. Sci. **52**, 9424–9429 (2017)
156. S. Zhang, W. Qin, M. Liu, X. Ren, Guomiao, C. Yuan, L. Yang, S. Yin, Facile preparation of Ag–Ag2S hetero-dendrites with high visible light photocatalytic activity. J. Mater. Sci. 53:6482–6493, (2018)
157. Ze-Da Meng, Lei Zhu, Kefayat Ullah, Shu Ye, Qian Sun, Oh Won-Chun, Enhanced visible light photocatalytic activity of Ag2S-graphene/TiO2 nanocomposites made by sonochemical synthesis. Chin. J. Catal. **34**, 1527–1533 (2013)

158. H.H. Li, J. Yang, J.Y. Zhang, M. Zhou, RSC Adv. **2**, 6258–6261 (2012)
159. K. Mageshwari, R.H. Sathyamoorthy, Mater. Sci. Semicond. Process. **16**, 43–50 (2013)
160. F. Chen, Y. Cao, D. Jia, J. Colloid Interface Sci. **404**, 110–116 (2013)
161. F. Liu, X. Shao, S. Yang, Mater. Sci. Semicond. Process. **34**, 104–108 (2015)
162. J. Wang, X. Zhang, J. Wu, H. Chen, S. Sun, J. Bao, S. Li, H. Bi, Nanoscale (2017). https://doi.org/10.1039/C7NR04593J
163. S.V. Prabhakar Vattikuti, C. Byon, Bi_2S_3 nanorods embedded with MoS_2 nanosheets composite for photodegradation of phenol red under visible light irradiation, Superlattices Microstructures **100,** 514–525 (2016)
164. Z. Hosseinpour, S. Hosseinpour, Mater. Sci. Semiconductor Process
165. M. Basu, A.K. Sinha, M. Pradhan, S. Sarkar, Y. Negishi, T. Pal, Environ. Sci. Technol. **44**, 6313–6318 (2010)
166. Y. Xie, Z. Hosseinpour, A. Alemi, A. Khandar, X. Zhao, New. J. Chem. (2015). https://doi.org/10.1039/C4NJ02298J

Chapter 6
Hybrid Nanomaterials for Advanced Photocatalysis

6.1 Semiconductor/Semiconductor-Based Hybrid Heterojunctions

For enhanced visible-light absorption, inhibited recombination of electrons, and holes and higher photocatalytic efficiency, the designing of heterostructures is an advanced and influencing technique [1]. Heterostructure gives preferential band alignment to obtain visible-light active photocatalytic activity [2]. In this section, types of hybrid heterostructure including conventional heterostructures (type-I, type-II, type-III), *p-n* heterojunctions, Z-scheme heterojunctions, and S-scheme heterojunctions will be summarized.

6.1.1 Conventional Heterojunction Photocatalysts

Generally, heterojunction is developed between semiconductors having unlike band gaps and band potentials properties [3]. The main categories of typical heterostructure photocatalysts include type-I, type-II, and type-III as displayed in Fig. 6.1 [3]. As shown in Fig. 6.1a, in type-I heterostructure photocatalyst, the semiconductor-A has lower valence band (VB) while its conduction band is higher in energy compared to semiconductor-B valence and conduction bands [4]. It was seen that upon light exposure, the positive and negative charges could be stored at valence and conduction bands of semiconductor-B only, thus these charge carriers could not be separated efficiently. In type-II heterostructure, semiconductor-A has high energy conduction and valence bands than semiconductor-B (Fig. 6.1b). Therefore, upon light illumination photo-excited electrons will move to semiconductor-B, while the positive holes will transfer to semiconductor-A, causing an improved separation of charge carriers [5, 6].

M. B. Tahir and K. N. Riaz, *Nanomaterials and Photocatalysis in Chemistry*,
Materials Horizons: From Nature to Nanomaterials,
https://doi.org/10.1007/978-981-16-0646-5_6

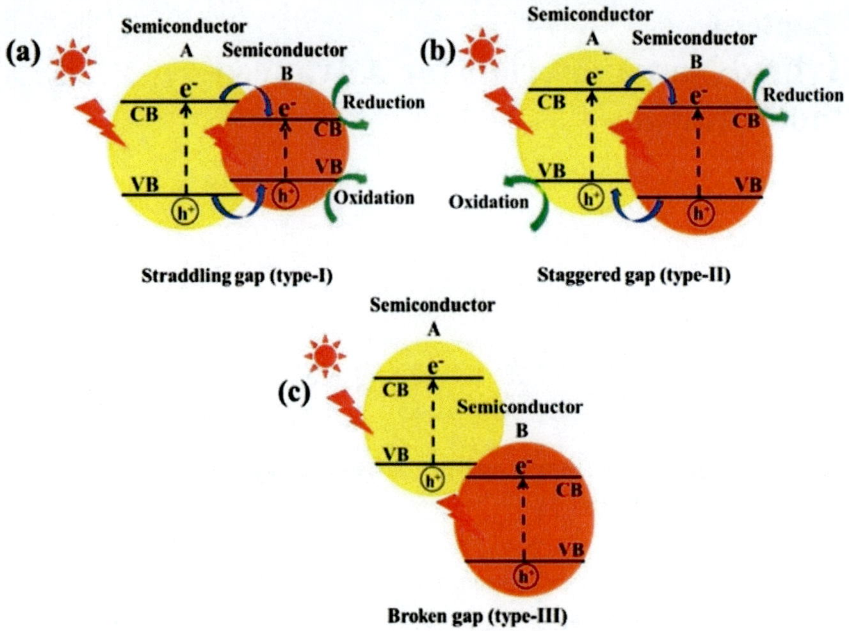

Fig. 6.1 Schematically illustrated diagram of three forms of separation mechanism of charge carriers in conventional light active heterostructures: **a** type-I, **b** type-II, and **c** type-III [3]

However, redox ability of type-II heterostructure will be decreased because reactions of reduction and oxidation occur on semiconductor-B having lesser reducing potential and on semiconductor-A of less oxidizing potential, respectively. Furthermore, the construction of type-III heterostructure material is same like type-II heterostructure photocatalyst excluding staggered gap which converts to wide gap due to which band gaps do not mix with each other [3, 7]. In this way, charge carriers transfer and separation cannot happen for type-III heterostructure, which make it unfavorable for improving the inhibition of charge carriers as well as photocatalytic performance. Therefore, only type-II heterostructure is the highly efficient among conventional heterojunctions to be used for enhancing the photocatalysis performance. In order to design type-II heterojunction photocatalysts, different combinations of semiconductors were utilized including g-C_3N_4/TiO_2, $BiVO_4$-WO_3, g-C_3N_4-WO_3, and CeO_2/g-C_3N_4 [8–11], etc., for increasing photocatalysis efficiency. For instance, Ma et al. [11] fabricated CeO_2/g-C_3N_4 nanosheets-based type-II heterostructure which showed 94% bisphenol A (BPA) degradation in 80 min under visible photo-illumination as compared to pristine CeO_2 (14%) and g-C_3N_4 (65%). The higher efficiency of photocatalyst was due to similar energy levels of CeO_2 and g-C_3N_4 which supported the transfer and separation of photo-excited electrons and holes. Therefore, photoproduced positive charges and superoxide radicals (O_2^-) enhanced photocatalytic elimination of bisphenol A. In another report, Wetchakun et al. used hydrothermal method for the preparation of $BiVO_4$/CeO_2

type-II heterostructure photocatalyst and applied for removal of both methyl orange and methylene blue dyes [12]. They observed that the isoelectric point difference of both semiconductors played an effective role in absorbing both cationic and anionic dyes at the same time and finally enhanced the efficiency of hybrid photocatalyst. Beside good charge carrier's separation efficiency of type-II heterostructure nanomaterials, their practical applications are limited due to some other problems. Therefore, formation of more efficient heterostructures of photocatalytic materials is required urgently.

6.1.2 P-N *Heterojunction Photocatalysts*

The superior electron–hole pairs separation performance achieved with a type-II heterostructure is not satisfactory to inhibit fast recombination of charge carriers during photocatalysis. Therefore, a p–n heterostructure-based nanomaterial is a best choice to further improve the photocatalytic efficiency [13]. The *p-n* heterojunction-based photocatalysts could be designed via linking *p*-type and *n*-type semiconducting materials. The major advantage of *p-n* heterojunction photocatalyst is that it can produce an additional electric filed to enhance the charge transport for increasing photocatalysis performance [14]. Before light exposure, the electrons stored on the surface of *n*-type semiconducting material near *p-n* interface trying to move into *p*-type semiconducting material, leaving behind positive hole in *n*-type semiconductor [15–17]. Similarly, holes in the *p*-type semiconducting material close to *p-n* heterojunction try to transfer into *n*-type semiconducting material, which leaves behind negative charges. The transportation of charges will remain to continue until fermi-level equilibrium of heterojunction is obtained [18, 19]. Thus, an induced electric field is established at heterojunction between two semiconductors.

In a report, Zhang et al. [20] fabricated NiS-CdS nanorods-based p–n heterojunctions which showed enhanced photocatalytic hydrogen evolution performance. The development of p–n heterostructures facilitated the charge transport between two semiconducting materials and suppressed the recombination of electrons and holes. Therefore, H_2 generation rate through photocatalysis over NiS-CdS nanorods with p–n heterojunctions was 56.6 μmol h^{-1} as compared to pristine CdS (2.8 μmol h^{-1}). A *p-n* heterojunction of Ag_2O/g-C_3N_4 was synthesized via facile chemical precipitation technique [21]. The higher performance of heterostructure toward organic pollutants degradation was ascribed to the better separation of photocreated charges, extended visible-light absorption, and SPR of silver metal. In another report, Ahamed et al. [22] designed a *n*-TiO_2/*p*-CuO thin film heterostructure and applied for hazardous organic contaminants and also for metal ion (Ni^{2+}) reduction. The superior performance of fabricated heterostructure was due to development of heterostructure which enhanced charge separation and decreased recombination of photocreated charge carriers as shown in mechanistic diagram of Fig. 6.2.

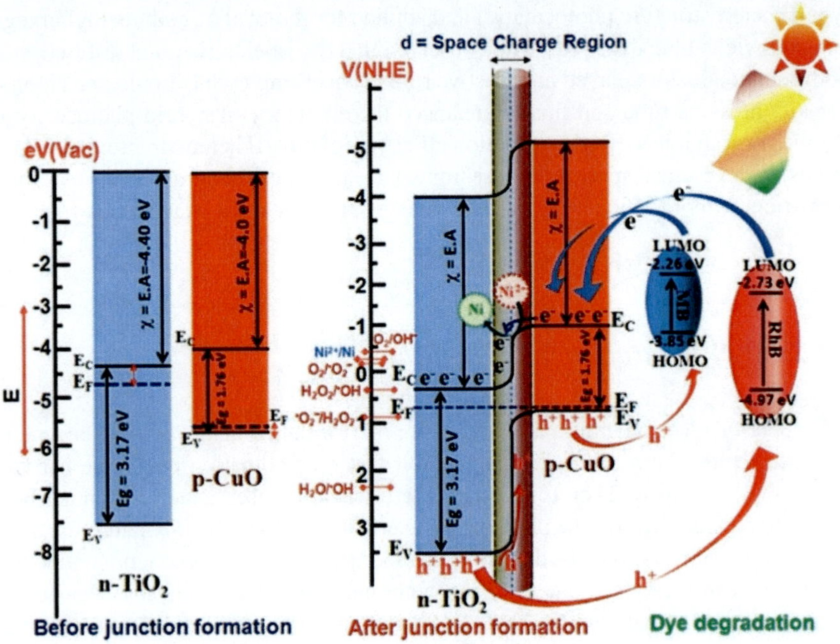

Fig. 6.2 Energy band representation of TiO_2/CuO heterostructure, before and after contact formation, with photocatalysis mechanism for disintegration of organic dyes and Ni^{2+}ion reduction process [22]

6.1.3 Z-Scheme Heterojunction Photocatalysts

6.1.3.1 Traditional Z-Scheme Heterojunction Photocatalysts

In order to increase the redox ability of nanomaterials, traditional Z-scheme photocatalytic materials have been firstly introduced by Bard in 1979 [23]. These photocatalysts showed higher charge separation as well as redox ability in photocatalysis reactions. This type of heterojunction system is developed using two semiconductors with an appropriate transitional couples, including Fe^{3+}/Fe^{2+}, IO_3^-/I^-, I_3^-/I^-. In both these semiconductors, the structure has staggered band alignment [24, 25]. Ideally in this heterojunction system, photo-induced positive charge carriers in valence band of photocatalytic material I (PS I) make reaction with electron donors (D), forming electron acceptors (A), while the photo-excited electrons in conduction band of photocatalytic material II (PS II) make reaction with A, producing D. Next, the reserved photocreated negative charges in conduction band of PS I and positive charge carriers in valence band of PS II take part in reducing and oxidizing reactions, respectively, as shown in Fig. 6.3 [3, 26]. Due to this, the system will achieve strong ability along with detached redox reaction active sites. However, this heterojunction structure has many drawbacks, for example, the shuttle redox ion pairs are necessary to support

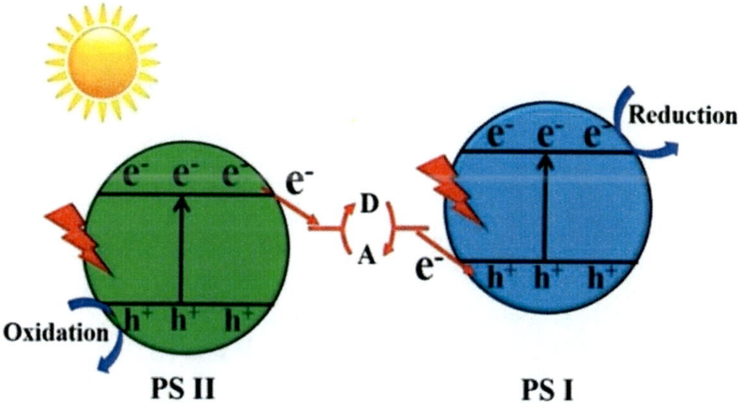

Fig. 6.3 Schematically illustrated scheme of charge separation mechanism on the conventional Z-scheme photocatalysis structure under photo-illumination [3]

transportation of photogenerated charges in traditional Z-scheme heterostructure and these ions can obtain enough transfer rates in only solution. Therefore, the solution phase requirement restricts the practical uses of traditional Z-scheme heterojunction.

6.1.3.2 All-Solid-State Z-Scheme Heterojunction Photocatalysts

In 2006, all-solid-state Z-scheme heterostructure was firstly proposed to get desired charge transport and to increase scope of applications [27]. This type of Z-scheme heterostructure consists of two semiconductor photocatalysts and a solid electron mediator between them. Upon light illumination, photogenerated electrons on PS II transfer to VB of the PS I through an electron mediator (including Pt, Au and Ag) and are further moved to CB of PS I as shown in Fig. 6.4 [3, 28]. Due to this, photo-induced electrons and holes are stored in the PS II having high oxidizing potential and in PS I with a greater reducing potential, respectively. This results into spatial electrons and holes separation along with optimized redox potential. Furthermore, all-solid-state Z-scheme-based heterojunction could be utilized in liquid, gas, and solid phases, which extend its practical applications [1].

Lan et al. [29] successfully prepared all-solid-state Z-scheme Bi@β-Bi_2O_3/g-C_3N_4 heterostructure using deposition and oxidation process, where Bi was applied as an electron mediator. The Z-scheme heterojunction showed higher decomposition ability for 2,3-dihydroxynaphthalene (2,3-DHN) with a degradation percentage of 87% within 100 min time of visible-light illumination. The improved photocatalysis performance was due to Z-scheme heterostructure, supporting superior separation of photogenerated charges. Besides this excellent performance, the electron mediators required for transport of electrons and holes between two materials in all-solid-state Z-scheme photocatalytic materials are costly and rare, which restricts its commercial scale use.

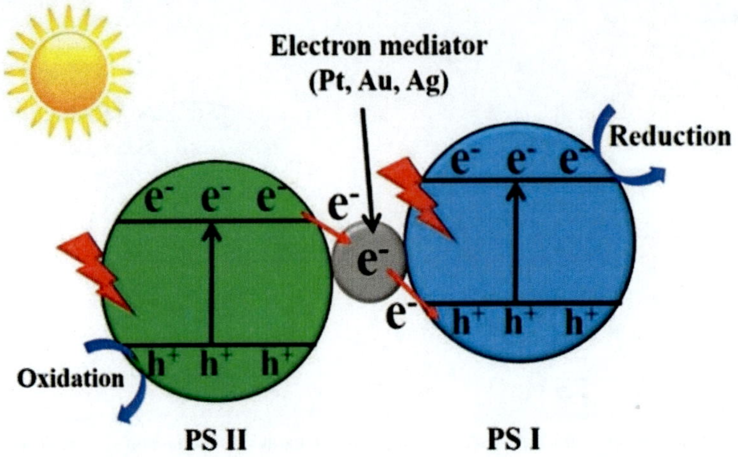

Fig. 6.4 Schematically represented diagram of charge separation using all-solid-state Z-scheme photocatalysts under photo-illumination [3]

6.1.3.3 Direct Z-Scheme Heterojunction Photocatalysts

The idea of direct Z-scheme heterojunction photocatalyst was first time introduced by Yu et al. [30] in 2013. They prepared direct Z-scheme heterojunction photocatalytic material via mixing two semiconducting materials without the use electron mediator [3, 30]. It was observed that transfer mechanism of charges in direct Z-scheme heterostructure is highly effective compared to type-II heterostructure due to good electrostatic interaction between positive and negative charges. Jia et al. [31] prepared direct Z-scheme g-C_3N_4/TiO_2 heterostructure with good visible-light-driven photocatalysis activity for methylene blue removal. The development of heterojunction (Fig. 6.5) between two semiconducting materials decreased e^-/h^+ pairs reunion which ultimately improved photocatalysis performance for methylene blue disintegration. Moreover, the charge transportation was also studied using density function theory (DFT) [32]. The results confirmed that due to different work function of both materials, an induced electric field was established, facilitating the transfer of charges to increase separation of charges. Jin et al. synthesized hierarchical-structured CdS/WO_3 direct Z-scheme photocatalytic material and utilized for photocatalysis reduction of carbon dioxide [33].

They observed higher separation ability of CdS–WO_3 for charges as compared to pure CdS and WO_3 HSs. They proposed and attributed this enhancement in charge separation and photocatalysis activity to establishment of CdS–WO_3 direct Z-scheme heterostructure which promoted negative charge accommodation on CdS surface, favoring the multi-electron CO_2 production (Fig. 6.6).

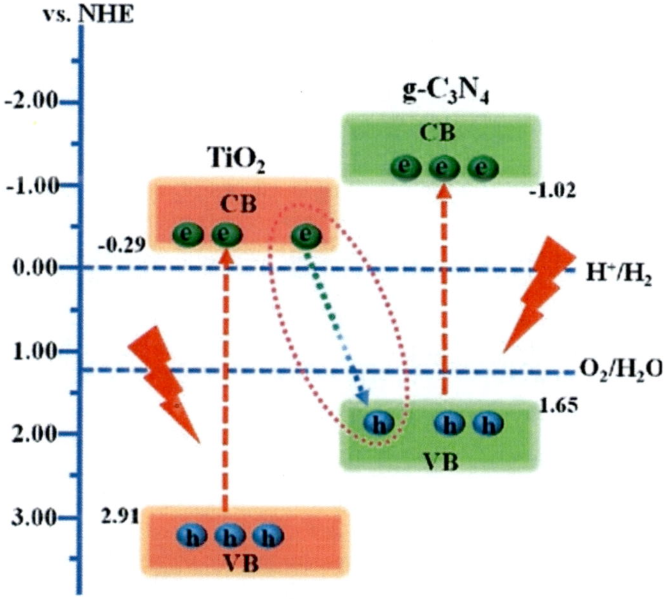

Fig. 6.5 Schematically illustrated mechanism describing MB removal over g-C_3N_4/TiO_2 heterostructure photocatalytic material when exposed to visible photo-illumination ($\lambda \geq 420$ nm) [31]

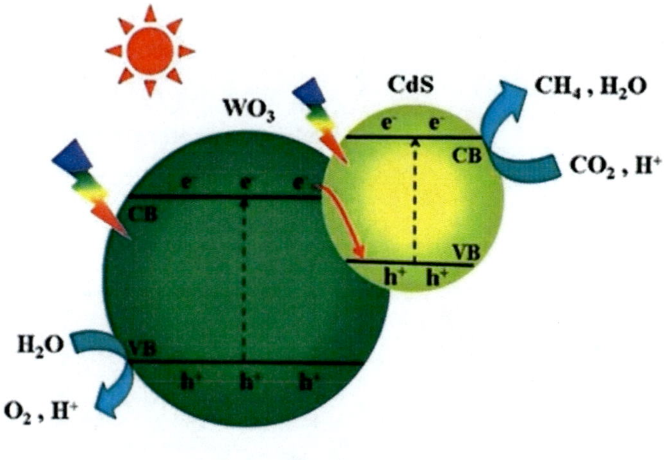

Fig. 6.6 Schematically illustrated charge transport mechanism for photo-excited charges on CdS/WO_3 direct Z-scheme heterostructure photocatalytic material when exposed to visible photo-illumination [33]

6.1.4 S-Scheme Heterojunction Photocatalysts

Recently, Yu et al. group has introduced the concept of S-scheme (or Step-scheme) heterojunction by modifying type-II heterostructure [34, 35]. Based on band structures classification, they sorted the photocatalysts into reduction photocatalysts (RPs) and oxidation photocatalysts (OPs) as displayed in Fig. 6.7a [35]. The reduction photocatalysts possessing higher conduction band are majorly used in solar fuels production. In RPs, the photo-induced electrons are useful, while photocreated holes are not effective and need to be removed from photocatalysis system via sacrificial agents. On the other hand, oxidation photocatalysts are mainly applied in environmental remediation, where photocreated holes play effective role while photocreated negative charges are useless. The S-scheme heterostructure consists of RP and OP with staggered band structures, which is likely to type-II heterostructure but with a

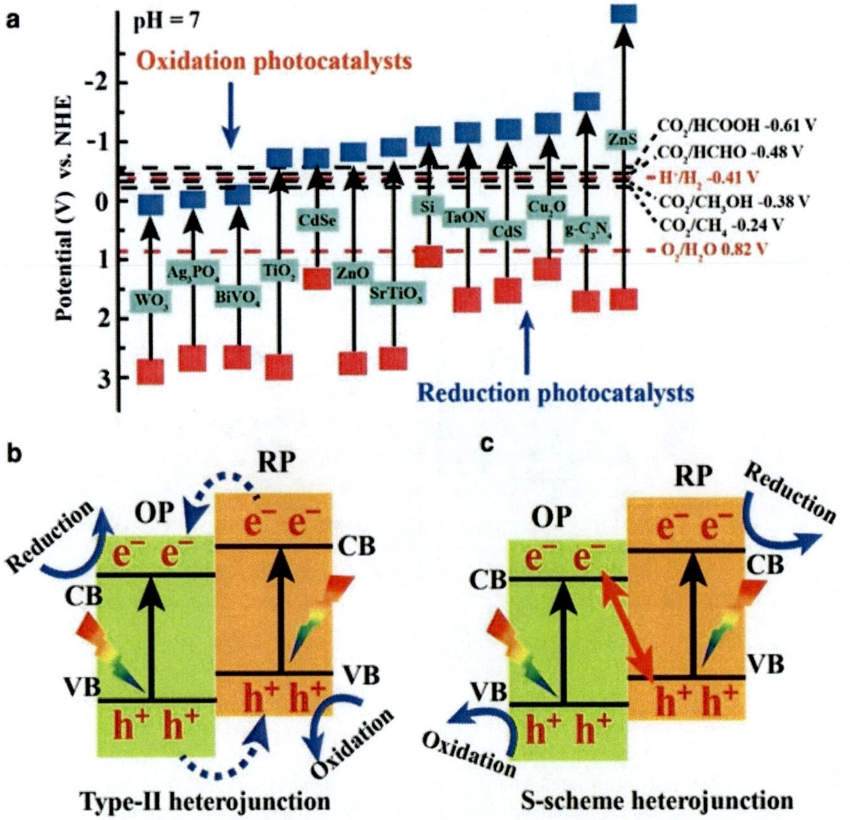

Fig. 6.7 Comparative analysis in charge transport between type-II and S-scheme heterostructures (**a**) band structures of some main photocatalytic materials. (**b** and **c**) band structure formation and charge transport pathway of (**b**) type-II heterostructure and (**c**) S-scheme heterostructure, reproduced from Ref. [35] with permission from Elsevier, copyright (2020)

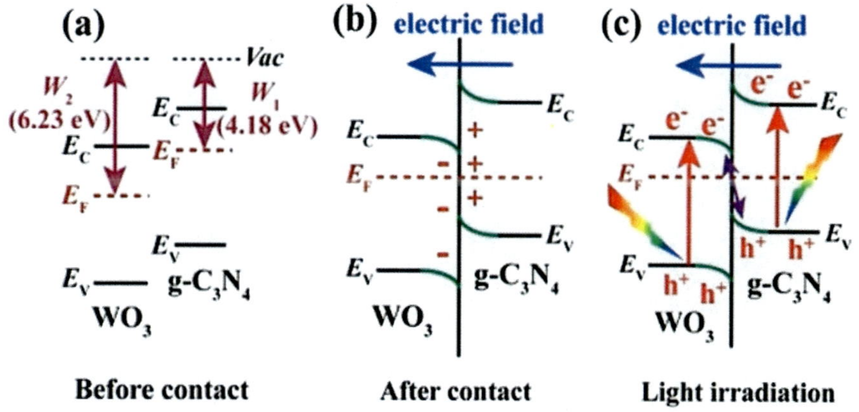

Fig. 6.8 **a** Representation of work functions of g-C_3N_4 and WO_3 materials. **b** the induced electric field and band edge bending at interconnection of WO_3/g-C_3N_4. **c** the S-scheme charge transport route between WO_3 and g-C_3N_4 when exposed to light irradiation, reproduced from Ref. [34] with permission from Elsevier, copyright (2019)

different charge transport pathway (Fig. 6.7b and c) [35]. It is known that in type-II heterostructure, the photo-induced charges are stored on the CB of OP and VB of RP, respectively, which results into weak redox ability. However, in S-scheme heterojunction, the high energy photo-induced electrons and holes are accumulated in the CB of RP and VB of OP, respectively, whereas the pointless photocreated positive and negative charges are recombined giving a higher redox potential [36–41]. For example, S-scheme heterojunction displayed significant improved spatial charge separation in case of WO_3/g-C_3N_4 S-scheme heterostructure with 2D and 2D surface contact. The prepared hybrid composite showed best performance for hydrogen evolution because the work function of tungsten trioxide is greater compared to g-C_3N_4 (Fig. 6.8a), which helped to induce induced electric field pointing from g-C_3N_4 to WO_3 and also band edge bending at their interconnection because of electron rearrangement (Fig. 6.8b). Due to this induced electric field, the photo-excited electrons in the CB of g-C_3N_4 and photo-induced holes in the VB of WO_3 with good redox ability are stored to take part in photocatalysis, whereas photo-induced positive charges in VB of g-C_3N_4 try to recombine with the photocreated negative charges in CB of tungsten trioxide (Fig. 6.8c).

6.2 Plasmonic Metal/Semiconductor-Based Hybrid Heterojunctions

The development of hybrid heterojunction based on plasmonic metal/semiconductor photocatalyst is an important strategy for extended visible-light absorption, improved charge carrier separation as well photocatalytic performance [42]. Various studies

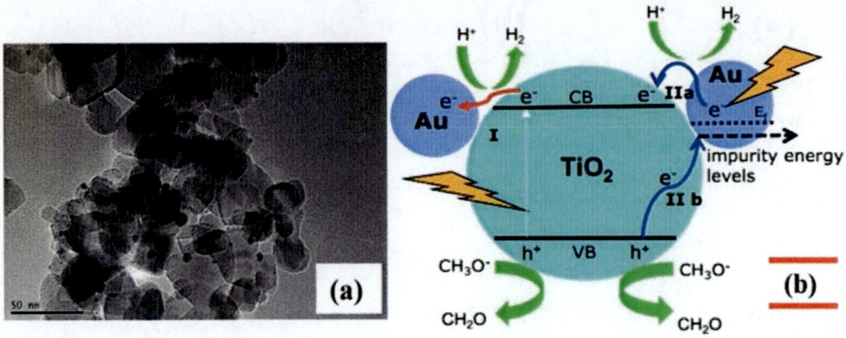

Fig. 6.9 **a** TEM image of Au (1 wt%)-TiO_2, **b** proposed reaction mechanism for hydrogen evolution over Au–TiO_2 in water/methanol mix solution under visible photo-illumination ($\geq$ 420 nm: process I + II; $\geq$ 500 nm: process II) [50]

have reported to investigate the influence of plasmonic metals such as Au and Ag on semiconductor photocatalysis under light illumination [43–46]. In early reports, researchers have used methanol as sacrificial agents to capture photo-excited holes to further allow the electrons to start some reduction reactions, whereas the role of noble metal was only to separate electron–hole pairs [43]. However, recently, plasmonic effect of noble metal especially Au nanoparticles and silver nanoparticles has attained a lot of attention because plasmon could promote the redox reaction via following pathways, enhancing light absorption, increasing charge separation, hot electron injection, and plasmon-induced energy transfer [47–49]. Nie et al. [50] reported Au loaded TiO_2 (Au-TiO_2) as shown in TEM image of Fig. 6.9a for photocatalytic H_2 evolution under visible-light irradiation. The results showed that Au could transfer the photogenerated electrons to the conduction band of TiO_2, and the Au-SPR effect is also beneficial to start electron–holepair creation (interfacial charge transfer process) upon visible-light irradiation ($\geq$500 nm) as displayed in the proposed mechanism of Fig. 6.9b. Furthermore, DFT calculation verified that Au-SPR could create new impurity energy levels within band gap of TiO_2.

In another study, Qiao et al. [51] used simple method to prepare Ag/mesoporous black TiO_2 nanotubes heterostructures (Ag-MBTHs) as presented in Fig. 6.10a and b. Due to Ti^{3+} self-doping and SPR effect of silver nanoparticles, the hybrid heterojunction showed superior visible NIR active photocatalysis efficiency for decomposition of 4-nitrophenol (4-NP) and removal of highly poisonous phenol. The authors attributed the enhanced performance of Ag-MBTHs to SPR effect of Ag, Ti^{3+} self-doping and heterostructure formation and good separation of photocreated charges (Fig. 6.10c).

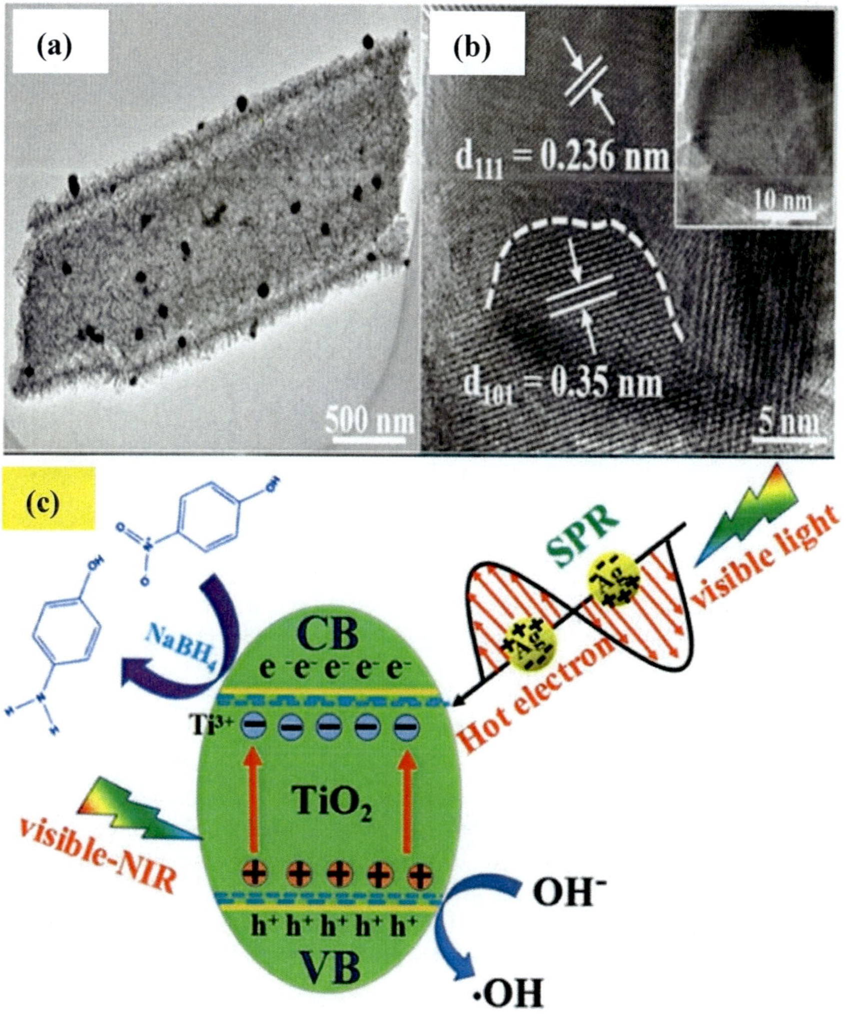

Fig. 6.10 **a** TEM, **b** HRTEM images of Ag/black TiO_2 heterostructure, **c** schematically representation of visible NIR active photocatalysis and photothermal catalytic reaction mechanism for Ag-MBTHs [51]

6.3 Semiconductor/Graphene-Based Hybrid Heterojunctions

Graphene is a two-dimensional (2D) single layer sheet of sp^2-hybridized carbon atoms with hexagonally packed structure, which has excellent physical characteristics such as superior charge accepting and transfer properties, good optical properties, higher thermal conductivity, large theoretical specific surface area, and good mechanical strength [52–54]. A lot of efforts were made to couple graphene with other semiconductors for the fabrication of heterojunctions photocatalysts with enhanced photocatalytic efficiency [3]. The good electron conducting nature of graphene permits the transfer of negative charges from semiconducting material to its surface, improving electron–hole pairs separation. For example, Zhang et al. [55] showed that photocatalytic performance of P25 (TiO_2) was really improved due to loading graphene for the degradation of methylene blue (MB) dye. They ascribed the enhanced activity to the fast transportation of electrons across the P25-graphene heterojunction and number of surface active sites because of higher specific surface area of graphene in hybrid nanomaterial. Recently, Beura et al. [56] prepared Ag-decorated ZnO-graphene hybrid photocatalyst via hydrothermal method. The morphological results indicate that Ag loaded ZnO nanorods were successfully embedded within the sheets of graphene (Fig. 6.11a–e) along with purity of nanohybrid which contains only Zn, O, C, and Ag elements (Fig. 6.11f). Furthermore, the nanohybrid showed good activity for degradation of methyl orange, and its degradation mechanism was described via band diagram (Fig. 6.11g). The authors attributed the enhanced efficiency of nanohybrid to extended visible-light absorption and decreased recombination of charges.

In a report, Yu et al. fabricated CdS nanorods/RGO hybrid composite and applied for CO_2 reduction [57]. The monohybrid composite showed higher adsorption ability and specific surface area due to excellent properties of graphene. Furthermore, the π–π conjugation interface between graphene and CO_2 molecules also facilitated the reduction of CO_2 into methane (CH_4). Therefore, the photocatalytic performance of CdS nanorods/RGO hybrid toward CH_4 production was superior as compared to pure CdS nanorods because of better charge carrier's separation efficiency, higher CO_2 adsorption, and good carbon dioxide conversion.

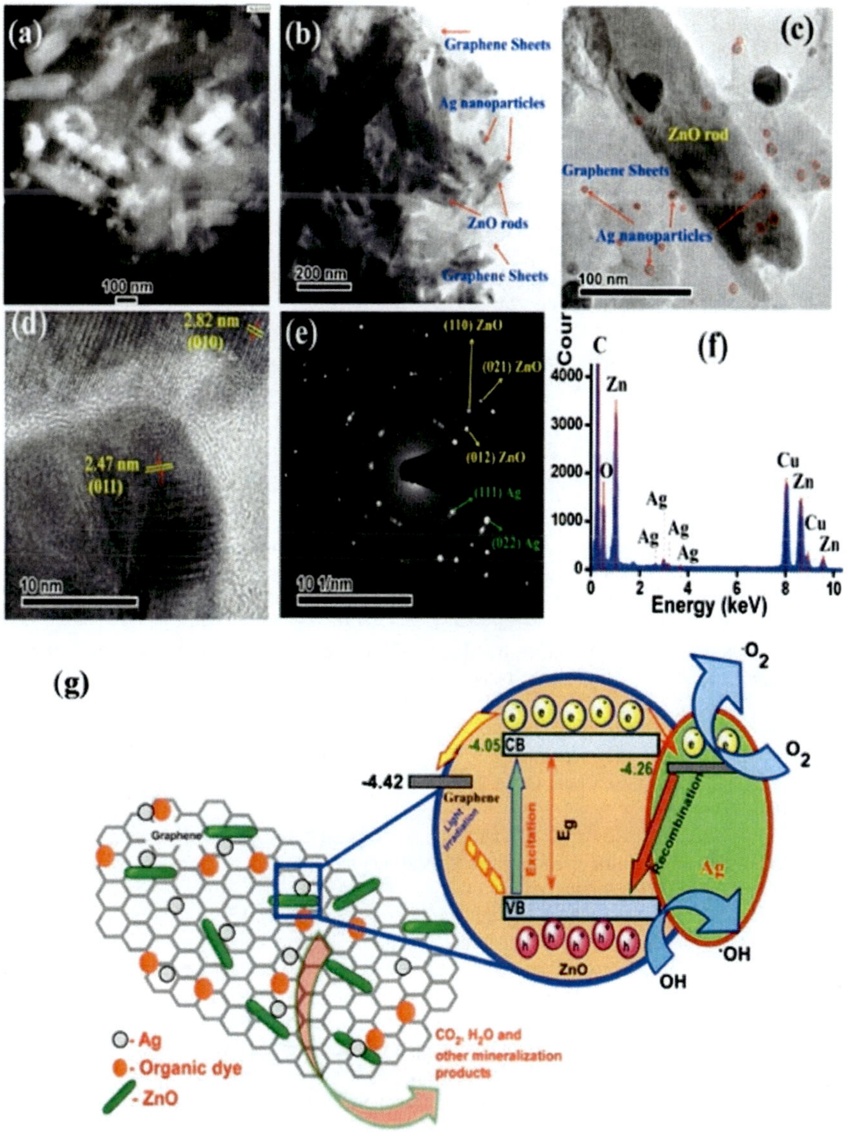

Fig. 6.11 TEM images of GZAg with 3 wt% silver doping: **a** HAADF-STEM, **b**, **c** bright field, **d** high-resolution image, **e** SAED pattern, and **f** XEDS spectrum, **g** schematically illustrated diagram which represents most probable mechanisms in photocatalysis disintegration of methyl orange using a GZAg composite as photocatalytic material, reproduced from Ref. [56] with permission from Elsevier, copyright (2021)

References

1. A.B. Djurišic, Y. He, A.M.C. Ng, Visible-light photocatalysts: Prospects and challenges. APL Mater. **8**, 030903 (2020)
2. K. Afroz, Md. Moniruddin, N. Bakranov, S. Kudaibergenovbc, N. Nuraje, A heterojunction strategy to improve the visible light sensitive water splitting performance of photocatalytic materials. J. Mater. Chem. A **6**, 21696 (2018)
3. J. Low, J. Yu, M. Jaroniec, S. Wageh, A.A. Al-Ghamdi, Heterojunction photocatalysts. Adv. Mater. 1601694 (2017)
4. H. Zhou, Y. Qu, T. Zeid, X. Duan, Energy Environ. Sci. **5**, 6732 (2012)
5. R. Marschall, Adv. Funct. Mater. **24**, 2421 (2014)
6. S.J. Moniz, S.A. Shevlin, D.J. Martin, Z.-X. Guo, J. Tang, Energy Environ. Sci. **8**, 731 (2015)
7. W. Shi, N. Chopra, Nanomater. Energy **2**, 158 (2013)
8. K. Li, S. Gao, Q. Wang, H. Xu, Z. Wang, B. Huang, Y. Dai, J. Lu, A.C.S. Appl, Mater. Interfaces **7**, 9023 (2015)
9. S.J. Hong, S. Lee, J.S. Jang, J.S. Lee, Energy Environ. Sci. **4**, 1781 (2011)
10. L.Y. Huang, H. Xu, Y.P. Li, H.M. Li, X.N. Cheng, J.X. Xia, Y.G. Xu, G.B. Cai, Dalton Trans. **42**, 8606 (2013)
11. R. Ma, S. Zhang, L. Li, P. Gu, T. Wen, A. Khan, S. Li, B. Li, S. Wang, X. Wang, Enhanced visible-light-induced photoactivity of type-II $CeO_2/g\text{-}C_3N_4$ nanosheet towards organic pollutants degradation. ACS Sustain. Chem. Eng. **7**, 9699–9708 (2019)
12. N. Wetchakun, S. Chaiwichain, B. Inceesungvorn, K. Pingmuang, S. Phanichphant, A.I. Minett, J. Chen, A.C.S. Appl, Mater. Interfaces **4**, 3718 (2012)
13. Y.-P. Yuan, L.-W. Ruan, J. Barber, S.C.J. Loo, C. Xue, Energy Environ. Sci. **7**, 3934 (2014)
14. L. Li, P.A. Salvador, G.S. Rohrer, Nanoscale **6**, 24 (2014)
15. L. Ye, J. Liu, L. Tian, T. Peng, L. Zan, Appl. Catal. B. 134–135, 60 (2013)
16. Z. He, Y. Shi, C. Gao, L. Wen, J. Chen, S. Song, J. Phys. Chem. C **118**, 389 (2014)
17. J.Q. Liu, L.L. Ruan, S.B. Adeloju, Y.C. Wu, Dalton Trans. **43**, 1706 (2014)
18. Y. Peng, M. Yan, Q.-G. Chen, C.-M. Fan, H.-Y. Zhou, A.-W. Xu, J. Mater. Chem. A **2**, 8517 (2014)
19. P. Murugesan, J.A. Moses, C. Anandharamakrishnan, Photocatalytic disinfection efficiency of 2D structure graphitic carbon nitride-based nanocomposites: a review. J. Mater. Sci. **54**, 12206–12235 (2019)
20. J. Zhang, S.Z. Qiao, L.F. Qi, J.G. Yu, Phys. Chem. Chem. Phys. **15**, 12088 (2013)
21. S. Liang, D. Zhang, X. Pu et al., Separation and purification technology a novel $Ag_2O/g\text{-}C_3N_4$ p–n heterojunction photocatalysts with enhanced visible and near-infrared light activity. Sep. Purif. Technol. **210**, 786–797 (2019)
22. S.T. Ahamed, A. Ghosh, B. Show, A. Mondal, Fabrication of n-TiO_2/p-CuO thin-film heterojunction for efficient photocatalytic degradation of toxic organic dyes and reduction of metal ions in solution. J. Mater. Sci. Mater. Electron. https://doi.org/10.1007/s10854-020-04217-6
23. A.J. Bard, Photoelectrochemistry and heterogenous photo-catalysis at semiconductors. J. Photochem. **10**, 59–75 (1979)
24. P. Zhou, J. Yu, M. Jaroniec, All solid-state Z-scheme photocatalytic systems. Adv. Mater. **26**, 4920–4935 (2014)
25. H. Li, W. Tu, Y. Zhou, Z. Zou, Z-scheme photocatalytic systems for promoting photocatalytic performance: recent progress and future challenges. Adv. Sci. **3**, 1500389 (2016)
26. K. Maeda, Z-scheme water splitting using two different semiconductor photocatalysts. ACS Catal. **3**, 1486–1503 (2013)
27. H. Tada, T. Mitsui, T. Kiyonaga, T. Akita, K. Tanaka, All-solid-state Z-scheme in CdS–Au–TiO_2 three-component nanojunction system. Nat. Mater. **5**, 782–786 (2006)
28. L.B. Jiang, X.Z. Yuan, G.M. Zeng, J. Liang, Z.B. Wu, H. Wang, Environ. Sci. Nano **5**, 599–615 (2018)

29. Y. Lan, Z. Li, D. Li, W. Xie, G. Yan, S. Guo, Visible-light responsive Z-scheme Bi@βBi$_2$O$_3$/g-C$_3$N$_4$ heterojunction for efficient photocatalytic degradation of 2,3-dihydroxynaphthalene. Chem. Eng. J. (2019). https://doi.org/10.1016/j.cej.2019.123686
30. J.G. Yu, S.H. Wang, J.X. Low, W. Xiao, Phys. Chem. Chem. Phys. **15**, 16883 (2013)
31. J. Jia, Y. W, Xu Meiling, M. Qi, W. Yanling, G. Zhao, MOF-derived the direct Z-scheme g-C$_3$N$_4$/TiO$_2$ with enhanced visible photocatalytic activity. J. Sol-Gel. Sci. Technol. **93**, 123–130 (2020)
32. J. Liu, B. Cheng, J. Yu, A new understanding of the photocatalytic mechanism of the direct Z-scheme g-C$_3$N$_4$/TiO$_2$ heterostructure. Phys. Chem. Chem. Phys. **18**, 31175–31183 (2016)
33. J. Jin, J.G. Yu, D.P. Guo, C. Cui, W.K. Ho, Small **11**, 5262 (2015)
34. J. Fu, Q. Xu, J. Low, C. Jiang, J. Yu, Ultrathin 2D/2D WO$_3$/g-C$_3$N$_4$ steps cheme H$_2$-production photocatalyst. Appl. Catal. B **243**, 556–565 (2019)
35. X. Quanlong, L. Zhang, B. Cheng, J. Fan, Y. Jiaguo, S-scheme heterojunction photocatalyst. Chem (2020). https://doi.org/10.1016/j.chempr.2020.06.010
36. F. Mei, K. Dai, J. Zhang, W. Li, C. Liang, Construction of Ag SPR-promoted stepscheme porous g-C$_3$N$_4$/Ag$_3$VO$_4$ heterojunction for improving photocatalytic activity. Appl. Surf. Sci. **488**, 151–160 (2019)
37. X. Jia, Q. Han, M. Zheng, H. Bi, One pot milling route to fabricate step-scheme AgI/I-BiOAc photocatalyst: energy band structure optimized by the formation of solid solution. Appl. Surf. Sci. **489**, 409–419 (2019)
38. J. Wang, Q. Zhang, F. Deng, X. Luo, D.D. Dionysiou, Rapid toxicity elimination of organic pollutants by the photocatalysis of environment-friendly and magnetically recoverable step-scheme SnFe$_2$O$_4$/ZnFe$_2$O$_4$ nano-heterojunctions. Chem. Eng. J. **379**, 122264 (2020)
39. Q. Xu, D. Ma, S. Yang, Z. Tian, B. Cheng, J. Fan, Novel g-C$_3$N$_4$/g-C$_3$N$_4$ S-scheme isotype heterojunction for improved photocatalytic hydrogen generation. Appl. Surf. Sci. **495**, 143555 (2019)
40. Q. Xu, L. Zhang, J. Yu, S. Wageh, A.A. Al-Ghamdi, M. Jaroniec, Direct Z-scheme photocatalysts: principles, synthesis, and applications. Mater. Today **21**, 1042–1063 (2018)
41. T. Di, Q. Xu, W. Ho, H. Tang, Q. Xiang, J. Yu, Review on metal sulphide-based Z-scheme photocatalysts. ChemCatChem **11**, 1394–1411 (2019)
42. L. Ma, S. Chen, Y. Shao, Y.-L. Chen, M.-X. Liu, H.-X. Li, Y.-L. Mao, S.-J. Ding, Recent progress in constructing plasmonic metal/semiconductor hetero-nanostructures for improved photocatalysis. Catalysts **8**, 634 (2018). https://doi.org/10.3390/catal8120634
43. G.L. Chiarello, E. Selli, L. Forni, Photocatalytic hydrogen production over flame spray pyrolysis-synthesised TiO$_2$ and Au/TiO$_2$. Appl. Catal. B: Environ. **84**, 332–339 (2008). [80] I.M. Arabatzis, T. Stergiopoulos, D. Andreeva, S. Kitova, S.G. Neophytides, P. Falaras, Characterization and photocatalytic activity of Au/TiO2 thin films for azo-dye degradation. J. Catal. **220**, 127–135 (2003)
44. R. Georgekutty, M.K. Seery, S.C. Pillai, A highly efficient Ag-ZnO photocatalyst: synthesis, properties, and mechanism. J. Phys. Chem. C **112**, 13563–13570 (2008)
45. W. Lu, S. Gao, J. Wang, One-pot synthesis of Ag/ZnO self-assembled 3D hollow microspheres with enhanced photocatalytic performance. J. Phys. Chem. C **112**, 16792–16800 (2008)
46. X.-G. Hou, M.-D. Huang, X.-L. Wu, A.-D. Liu, Preparation and studies of photocatalytic silver-loaded TiO$_2$ films by hybrid sol–gel method. Chem. Eng. J. **146**, 42–48 (2009)
47. W. Li, J.G. Valentine, Harvesting the loss: Surface plasmon-based hot electron photodetection. Nanophononics **6**, 177–199 (2017)
48. J. Li, S.K. Cushing, F. Meng, T.R. Senty, A.D. Bristow, N. Wu, Plasmon-induced resonance energy transfer for solar energy conversion. Nat. Photonics **9**, 601–607 (2015)
49. K. Wu, J. Chen, J.R. McBride, T. Lian, Efficient hot-electron transfer by a plasmon-induced interfacial charge-transfer transition. Science **349**, 632–635 (2015)
50. J. Nie, J. Schneider, F. Sieland, L. Zhou, S. Xia, D. Bahnemann, New insights into the surface plasmon resonance (SPR) driven photocatalytic H$_2$production of Au–TiO$_2$. RSC Adv. **8**, 25881–25887 (2018)

51. P. Qiao, B. Sun, H. Li, K. Pan, G. Tian, L. Wang, W. Zhou, Surface plasmon resonance-enhanced visible-NIR-driven photocatalytic and photothermal catalytic performance by Ag/Mesoporous Black TiO_2 nanotube heterojunctions. Chem. Asian J. https://doi.org/10.1002/asia.201801428
52. D. Chen, H. Zhang, Y. Liu, J. Li, Energy Environ. Sci. **6**, 1362 (2013)
53. H. Chang, H. Wu, Energy Environ. Sci. **6**, 3483 (2013)
54. N. Li, M. Cao, C. Hu, Nanoscale **4**, 6205 (2012)
55. H. Zhang, X.J. Lv, Y.M. Li, Y. Wang, J.H. Li, ACS Nano **4**, 380 (2010)
56. R. Beura, R. Pachaiappan, T. Paramasivam, Photocatalytic degradation studies of organic dyes over novel Ag-loaded ZnO-graphene hybrid nanocomposites. J. Phys. Chem. Solids **148**, 109689 (2021)
57. J.G. Yu, J. Jin, B. Cheng, M. Jaroniec, J. Mater. Chem. A **2**, 3407 (2014)

Chapter 7
Conclusion and Future Prospects

Today, the top global issues include guaranteed clean supply of water and ensured supply of clean and renewable energy at lower cost. For this, many researchers are investigating advanced nanomaterials and processes for water cleaning and production of renewable energy effectively at lower cost and with less energy consumption. Among different processes, the photocatalysis method using nanomaterials is an advanced technique which fulfills all requirements for both remediation of environmental pollutants as well as energy production. Therefore, the main focus of this book is on the basic photocatalytic processes and also on the development of nanomaterials used in photocatalysis for environmental remediation and energy production. For example, Chap. 1 describes the brief history of photocatalysis, broad definition of photocatalysis, and classification of photocatalysis. The photocatalysis is divided into two main categories including energy conversion (H_2 production, CO_2 reduction, and N_2 fixation) and environmental remediation (air purification and water purification).

Chapter 2 of this book highlights the fundamentals of photocatalysis for energy conversion such as hydrogen evolution from water splitting, CO_2 reduction into valuable fuels and N_2 fixation for ammonia synthesis with special prominence on their importance, basic principles, and performance evaluation methods. The nanomaterials as photocatalyst used for photocatalytic energy conversion applications are reviewed and discussed in Chap. 4. The fundamental mechanisms and principles of photocatalysis for air purification (removal of VOCs and inorganic gases) and water cleaning (degradation of water pollutants including CECs, EDCs, pathogenic germs, and cyanotoxins) are presented in Chap. 3. The nanomaterials including metal oxide and metal sulfide-based photocatalysts for environmental remediation are

M. B. Tahir and K. N. Riaz, *Nanomaterials and Photocatalysis in Chemistry*,
Materials Horizons: From Nature to Nanomaterials,
https://doi.org/10.1007/978-981-16-0646-5_7